Peter Topalovic, Gail Krantzberg (Eds.)
Responsible Care®

Also of Interest

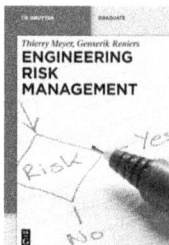

Thierry Meyer, Genserik Reniers
Engineering Risk Management, 2013
ISBN 978-3-11-028515-4, e-ISBN 978-3-11-028516-1

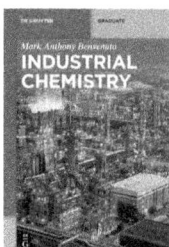

Mark Anthony Benvenuto
Industrial Chemistry, 2013
ISBN 978-3-11-029589-4, e-ISBN 978-3-11-029590-0

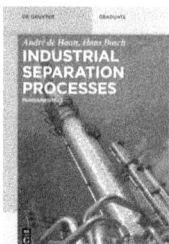

André de Haan, Hans Bosch
Industrial Separation Process, 2013
ISBN 978-3-11-030669-9, e-ISBN 978-3-11-030672-9

International Union of Pure and Applied Chemistry (IUPAC)
Pure and Applied Chemistry
ISSN 0033-4545, e-ISSN 1365-3075

International Union of Pure and Applied Chemistry (IUPAC)
Chemistry International
ISSN 0193-6484, e-ISSN 1365-2192

Peter Topalovic, Gail Krantzberg (Eds.)

Responsible Care®

A Case Study

Edited by Peter Topalovic and Gail Krantzberg
Published in cooperation with the International Union
of Pure and Applied Chemistry (IUPAC)

DE GRUYTER

Editors

Peter Topalovic, M.Eng.
School of Engineering Practice
McMaster University
Ctr. of Engineering & Public Policy
1280 Main Street West
Hamilton, ON, L8S 4L7
Canada
e-mail: topalopj@mcmaster.ca

Dr. Gail Krantzberg
School of Engineering Practice
McMaster University
Ctr. of Engineering & Public Policy
1280 Main Street West
Hamilton, ON, L8S 4L7
Canada
e-mail: krantz@mcmaster.ca

ISBN 978-3-11-034292-5
e-ISBN 978-3-11-034316-8
Set-ISBN 978-3-11-034317-5

Library of Congress Cataloging-in-Publication Data
A CIP catalog record for this book has been applied for at the Library of Congress.

Bibliographic information published by the Deutsche Nationalbibliothek
The Deutsche Nationalbibliothek lists this publication in the Deutsche Nationalbibliografie;
detailed bibliographic data are available in the Internet at http://dnb.dnb.de.

© 2014 Walter de Gruyter GmbH, Berlin/Boston

Typesetting: Compuscript Ltd.
Printing and Binding: Hubert & Co. GmbH & Co. KG, Göttingen
Cover image: Thinkstock/iStock
∞ Printed on acid-free paper
Printed in Germany

www.degruyter.com

Preface

The Responsible Care®1 IUPAC COCI Project

The first inkling of what has become the leading health safety and environmental ethos of the worldwide chemical industry occurred in the late 1970s. This involved the writing of one-page terms of reference for the Canadian Chemical Producers Association (CCPA). It lead to a more detailed definition of the ethos, which was fully acted on after the Bhopal disaster on December 3, 1984, when more than 40 tons of methyl isocyanate gas leaked from a pesticide plant in Bhopal, India, immediately killing at least 3,800 people and causing significant morbidity and premature death for thousands more.

The above disaster raised awareness for the need of enforceable international standards for environmental safety and preventative strategies to avoid similar accidents and also to promote industrial preparedness. By 1986 the CCPA Responsible Care® (RC) process was underway, and every member who subscribed to RC was required to take action necessary to show that they were in compliance with its codes of practice and were to be verified by a panel of experts and non-experts associated with industry.

Currently the Responsible Care® process has been adopted by over 50 countries, which are guided by the Responsible Care® Charter of the International Council of Chemical Associations (ICCA). The CCPA, now called the Chemistry Industry Association of Canada (CIAC), has continued to improve the process and is completing its sixth round of its mandatory triennial verification of members.

The International Union of Pure and Applied Chemistry (IUPAC) has been defining nomenclature and developing new initiatives that assist the world of chemicals in many ways. One of its key goals is to build capacity for the safe use of chemicals in a variety of places and especially in the developing world. Since the early 1990s, IUPAC's Committee on Chemistry and Industry (COCI) has sponsored a Safety Training Program, whereby middle-management people from developing countries apply for and are selected to receive safety training at host companies in the developed world.

Although Responsible Care® has been active for over 25 years, it is not well understood in broader chemistry organizations, such as IUPAC, academia, many chemical societies, and those countries that have no chemical production but do use many chemicals in their own commerce.

I was asked to join COCI to bring my industry experience with Responsible Care® to develop IUPAC projects to help expand the understanding of what it is, how it works, and show some examples.

This book is the result of two IUPAC funded projects. The first allowed us to produce the first two chapters, which describe Responsible Care® in Canada and

---

1 Responsible Care® is a registered trademark of the Chemistry Industry Association of Canada.

show its development. The second chapter discusses whether adherence to RC in the international chemical supply chain could have prevented a serious loss of life in Haiti. The third chapter is a positive story about Responsible Care® implementation and continued development over 25 years at a small Canadian company. Finally, there is documentation from a workshop held at McMaster University, which can be used as a model for readers to use this book as an educational tool.

Without the leadership and long-term commitment of Jean Belanger, who was CEO of the CCPA, Responsible Care® would not be the success it is today. He also contributed much to the first chapter and is regarded the "Father of Responsible Care®". In a recent conversation he commented on the value of Responsible Care® and the challenge that the chemistry community faces:

"The pace of change and knowledge will increasingly test a public that desperately wants to trust someone who is trustworthy. All those involved with chemistry, therefore, must walk the talk and do it openly and unfailingly, because the public has a long memory. This will require the best minds available, but they will only come if they truly believe that they can be proud of what the chemistry industries and sciences stand for. The ongoing success of Responsible Care®, however, will only be guaranteed through the continuing commitment of the industry's current and future leaders, alongside the leading scientists."

A great deal of work has gone into producing this book. I want to acknowledge Dr. Gail Krantzberg and Mr. Peter Topalovic for their excellent work and patience. The project would not have occurred without the financial support of the IUPAC Committee on Chemistry and Industry and the IUPAC Project Committee.

Our task now is to take this book out into the chemistry community and beyond, and to use it to help continued improvement in the safe production and use of chemicals around the world. I believe the codes of practice could foster social responsibility in many sectors, learning from chemical industry leadership.

Dr. Bernard West, Toronto, September 2013.

Contents

List of Contributing Authors

Jean Bélanger
211 Wurtemburg Street
Apartment 912
Ottawa, ON, K1N 8R4
Canada

Gail Krantzberg
School of Engineering Practice
McMaster University
Ctr. of Engineering & Public Policy
1280 Main Street West
Hamilton, ON, L8S 4L7
Canada
e-mail: krantz@mcmaster.ca

Maria Topalovic
School of Engineering Practice
McMaster University
1280 Main Street West
Hamilton, ON, L8S 4L7
Canada
e-mail: maria.topalovic@gmail.com

Peter Topalovic
School of Engineering Practice
McMaster University
Ctr. of Engineering & Public Policy
1280 Main Street West
Hamilton, ON, L8S 4L7
Canada
e-mail: topalopj@mcmaster.ca

Joanne West
43 Eglinton Avenue East
Apartment 1101
Toronto, ON, M4P 1A2
Canada
e-mail: joanne.s.west@gmail.com

Jean Bélanger, Peter Topalovic, Gail Krantzberg,
and Joanne West

1 Responsible Care: History and development

Abstract: This analysis offers insight into the development and evolution of the principles of Responsible Care[1] with the intention of providing readers in chemistry-related industries and academia with the insight and the means to promote and implement Responsible Care principles in their own work, related to their particular set of circumstances.

We first focus on the key elements of Responsible Care developed by the Canadian Chemical Producers' Association. This follows with an illustration of the need to develop a set of elements for self-examination to be used by chemistry-associated readers, whether in industry or academia, to assess their own needs and Responsible Care's applicability.

While Responsible Care was first developed in Canada, it has expanded to the United States and over 50 other countries through their national chemical associations. Further development has taken place under the guardianship of the International Council of Chemical Associations. The Responsible Care Global Charter has been signed by the CEOs of over 90 major multinational companies. Ongoing development is enriched by the exchange of international experience and adaptation to meet evolving concerns and differing cultures. It is hoped that other national associations will add to the findings in this report by describing their own experiences of development and providing a sound base for the adoption of Responsible Care in other regions and in other chemistry-related groups worldwide [1].

1.1 Responsible Care project

Responsible Care (RC) represents a commitment on the part of the chemical industry to contribute to the betterment of society while minimizing any adverse environmental impacts and societal consequences. It is a culture that focuses on doing the right thing and demonstrating its commitment to public verification.

While the RC initiative is well known in the chemical industry, it is less well understood within the academic and other technical communities. The International Union of Pure and Applied Chemistry (IUPAC) Committee on Chemistry and Industry (COCI) has now approved an initiative to create a framework project [2] focused on the responsible application of chemistry at all stages, from research, through industrial production, to the ultimate use and disposal of the products. RC will form the basis for understanding what responsibilities chemists have in using, handling, and producing chemicals. The implementation of this initiative will comprise a linked series of projects, and this paper represents the first stage of implementation.

---

1 Responsible Care® is a registered trademark of the Chemistry Industry Association of Canada.

The goal of the RC Project is to build knowledge and capacity about the basis, methodology, and goals of RC and targets the following audiences:
- Junior leaders in academic, business, and government organizations in developing countries.
- Chemistry associations in developing countries.
- Supply-chain contacts in developing countries and their suppliers in the developed countries.
- Various groups in developed countries that will benefit from this knowledge, including universities and government research establishments.

An important goal of this broadening process is to provide feedback to the implementers of RC in order to assist in the continuous improvement of the initiative. The results of this initiative are meant to increase the awareness and the application of the RC ethic in institutions related to chemistry worldwide, resulting in safer and more sustainable ways of designing, developing, and using chemical products.

1.2 Key principles underpinning Responsible Care

The basic principles of RC in Canada grew out of intensifying concerns that the chemical industry was risking the loss of its public license to produce [3]. Over the past 30 years, pressures to regulate the chemical industry have increased, exacerbated at times by major incidents resulting in severe health and environmental impacts. The most salient example is that of the Union Carbide plant in Bohpal India, which exploded due to failing safety systems in a storage facility in 1984 and resulted in thousands of casualties [4]. The leaders of the Canadian chemical industry have encountered a series of crossroads since this time. They could have chosen a path that would have led to engaging regulators in case-by-case challenges that would have diminished the industry's capabilities to develop its economic potential; however, original RC leaders understood that the problem was much more complex and required a more progressive solution [5].

In accordance with the findings of public opinion polls, industry leaders recognized that the public's concern is one of trust [6, 7]. The industry leaders chose a path that is anchored in gaining the trust of the affected communities and society in general. Trust has become the key driver of RC, which cannot be imposed but rather requires the application of three fundamental principles forming the cornerstones of RC.

1.2.1 Doing the right thing

Traditionally, the industry had responded to the laws and regulations in existence and reacted defensively to the introduction of new regulation [4, 7]. However, trust requires a commitment to "do the right thing", regardless of legal obligation. In fact, visibly doing the right thing is an essential step in building trust and contributing to a company's social license to operate [8, 9]. "Doing the right thing" radically changed the industry's focus from regulatory compliance to being ethically driven, including being an advocate for regulation to drive continuous improvement. This focus on corporate social responsibility in a complex world with a multiplicity of stakeholders requires that companies engage in meaningful dialogue with interested parties, namely, peers in other companies, individuals representing plant communities, and government representatives, incorporating the following principles [10]:

- Active listening to truly understand the underlying sources of concerns.
- Accurate presentation of the risks involved in operations and products and the steps taken to minimize them.
- A visible effort to integrate inputs from interested parties into planning and implementation processes.
- A broad consensus that the benefits provided by the company outweigh the risk.

Placing RC within an ethical framework represents a radical departure from what could have been a simple environmental risk management process. RC, operating as an ethical mandate, can be embedded into all aspects of the corporate culture, separating it from a standard environmental management system which is usually concerned with issues of quality or meeting environmental standards [11]. As an ethic it requires a radical change in corporate culture, requiring the Chief Executive Officer (CEO) to enlist buy-in from all employees and to take account of their environmental performance in an integrated and balanced manner with their economic performance [12].

1.2.2 Being open and responsive to public concerns

In the past, the industry had presumed that chemical issues were too complex to be understood by the general public. Industry believed that it knew best how to handle products and processes safely, and that the public should just trust the industry [7]. However, trust must be earned and secretiveness leads to the perception that externally imposed rules are needed. Openness is the basis of accountability [13]. The public is the final arbiter as to whether responsible management initiatives are truly protecting the environment. In addition to the public, industry requires that responsible management principles are universally adhered to, and companies need to ensure that their efforts can stand up to external scrutiny.

Responding to public concerns through an ethical approach requires sensitivity to the changing nature of concerns over time. RC has been successful in this, as evidenced by its evolution to include verification, Community Advisory Panels (CAPs), and the sustainable principles of green chemistry. The most recent revision of the ethic includes preventive and green engineering principles [14]. To maintain public trust, the industry must engage in ongoing and visible responsiveness to what is currently of public and scientific concern. In the early 1980s, the issues dominating the public agenda were closely related to plant operations or movement of chemicals. Since then, concerns have broadened to focus on product stewardship, including the use of the chemicals, and are currently dominated by concerns relating to sustainability and climate change [15, 16].

1.2.3 Caring about products from cradle-to-grave-to-cradle again

Canadian chemical industry leaders recognize that caring about products and their potential impact on people and the environment should not stop at the plant gate [14, 17]. While customers have responsibilities associated with use and disposal, chemical companies' ethics must drive the companies to advise and help those customers in proper handling. It may require withholding products from customers who fail to practice due diligence in appropriate handling and consumer use; a reflection of true product stewardship [17]. This is an important consideration, since as far as the public is concerned, all chemical producers and distributors are vulnerable to being negatively perceived as the weakest link in the supply chain, regardless of whether or not the industry has control over the products at a particular moment in time [4, 7].

Comprehensiveness also allows the chemical industry to proactively approach and deal with public concerns in the most cost-effective manner, with fewer duplications and a better time horizon. In this way, the industry becomes a credible and positive player in the search for solutions, one that should be listened to and trusted. This helps change the perception of the industry as being a problem creator to becoming a problem solver.

1.3 History of Responsible Care

Prior to the creation of RC in 1985, the Canadian Chemical Producer's Association (CCPA) celebrated its role in Canadian society in 1973 by releasing a pamphlet entitled "Canada's Invisible Industry" [18]:

> "The Canadian industrial chemical industry is the invisible industry – the fulcrum of Canada's inverted economic pyramid ... it is no beggar ... it seeks no preferential treatment to the detriment of other Canadian industries, it only seeks the opportunity to serve Canadian interests as the foundation upon which all our other industries must build ...."

In 1979, the association commissioned a study to assess the public standing of the industry. The report [19] stated that, while invisibility had some advantages, it precluded the building of community support for the industry. Soon after, the industry faced a major train derailment in which a rail car explosion, due to a faulty wheel bearing, caused the release of styrene, toluene, propane, caustic soda, and chlorine into areas surrounding Mississauga [20]. This and other safety and environmental issues (such as the Bhopal disaster) began to accumulate domestically and internationally. The industry quickly became visible for all the wrong reasons.

In 1983, CCPA members were asked to voluntarily sign a statement of guiding principles of industry behavior [21]. Theses principles stated, in part, that companies would:

- Ensure their operations did not present an unacceptable level of risk to employees, customers, the public, or the environment.
- Provide relevant information on the hazards of chemicals to customers and the public.
- Be responsive and sensitive to legitimate community concerns.
- Proactively communicate to stakeholders a commitment to continuous improvement in environmental, health and safety performance.

In 1983, the Canadian federal government commissioned a study to determine the future direction of the Canadian chemical industry and the role of the government in regulating the industry. The study committee consisted of 12 chemical industry leaders and the leader of the major trade union. Initially it sought to demonstrate how crucial the chemical industry was to Canada's economy. However, the study also caused some of the participating CEOs on the project to question whether the general public also perceived the industry as integral to this economy or saw it as creating more problems than it was worth. This reflection renewed emphasis on the concept of responsible management of chemicals and a draft statement of guiding principles that had been developed five years earlier [22]. At the time of the draft, legal counsel expressed concerns that such a statement could lead to increased liabilities; however, eventually the draft statement was reviewed and approved by the Board of Directors of the CCPA.

The committee's report was published in February 1984 and signed by all project members. It affirmed that the industry was a key and responsible industry and published for the first time the Statement of Guiding Principles, citing the Transportation Emergency Assistance Program as a tangible example of the CCPA's commitment to implementing RC. The report stated that formal acceptance of the Guiding Principles by all members of the CCPA would be actively sought.

Months later, in 1984, the cataclysmic Bhopal disaster occurred and a special CCPA Board meeting convened immediately after the incident in December 1984. Although 96% of member companies had signed on to the inaugural version of RC, doing so was not mandatory. At the meeting, a motion to make it mandatory was approved unanimously. All companies were then urged to review their current safety practices to identify potential weaknesses and report their findings. At a special meeting in January 1985 members agreed that the industry would only be as good as its weakest link; any company taking its responsibilities lightly would negate efforts of the whole industry. RC was born.

This series of meetings illustrates the important evolution that took place in the minds of Canadian chemical industry leaders. They recognized a need to develop trust with the public. To meet that need, a committee was formed to develop a common Safety Assessment Process. A second committee was tasked with developing a set of codes governing responsible management across the life cycle of chemical products [23]. Today, this is an accepted concept within the industry [17], but in 1985 it was revolutionary. It involved concepts such as refusing to produce a product if one was uncertain it would be handled with due diligence, informing plant neighbors about the dangers associated with the processes inside the plant gates and the precautions taken to mitigate those [3]. Each Board member also agreed to visit three non-Board member CEO's to secure their commitment to RC and offer help in implementation.

In the early stages of RC Code development, founding members pointed out that legal requirements offered a blueprint for minimum action and turned the focus of RC to employ an ethical mandate that aimed to meet or exceed regulation, which was considered the minimum amount of action required. This became a crucial turning point in the history of RC. The CEOs on the Board determined what was needed in order to do the right thing and incorporated that philosophy into their mandate. Board members recognized that to meet the goal of developing public trust, they needed to incorporate a verification process into RC [23]. The result was the first voluntary publication of their quantified performance on individual chemical emissions and five-year projections of emissions reduction, which formed the basis of the National Emissions Reduction Masterplan, or NERM [24]. They also chose to publicize the National Advisory Panel's (NAP's) annual performance reports, which record emissions reductions of member companies and adoption rates of response care philosophies.

1.4 Basic philosophy

1.4.1 Building trust in Canada: An on-going mission

RC has developed as an ethically driven way of life, a culture reflecting a unique vision of member companies' corporate and social responsibilities. It addresses the reality that corporate values must emphasize long-term commitments to: community and occupational health and safety, minimizing environmental footprint, and promoting the mitigation of environment impacts across the industry. This must be done within a framework that minimizes social impacts and attempts to create positive relationships with all stakeholders.

The first Statement of Commitment to RC was published in February 1984 by the CCPA [25] stating:

> "The Canadian chemical industry is committed to taking every practical precaution to ensure that products do not present an unacceptable level of risk to its employees, customers, the public or the environment."

It further outlined a set of guiding principles:
- Ensure that its operations do not present an unacceptable level of risk to employees, customers, the public, and the environment.
- Provide relevant information on the hazards of chemicals to its customers, urging them to use and dispose of products in a safe manner, and make such information available to the public on request.

- Make RC an early and integral part of the planning process leading to new products, processes or plants.
- Increase the emphasis on the understanding of existing products and their uses and ensure that a high level of understanding of new products and their potential hazards is achieved prior to and throughout commercial development.
- Comply with all legal requirements which affect its operations and products.
- Be responsive and sensitive to legitimate community concerns.
- Work actively with and assist governments and selected organizations to foster and encourage equitable and attainable standards.

In 2003, the CCPA reviewed its commitment and revised the Statement of Commitment [26]: "We are committed to do the right thing and be seen to do the right thing. We are guided towards environmental, societal, and economic sustainability by the following principles:

- We are stewards of our products and services during their life cycles in order to protect people and the environment.
- We are accountable to the public, who have the right to understand the risks and benefits of what we do and to have their input heard.
- We respect all people.
- We work together to improve continuously.
- We work for effective laws and standards, and will meet or exceed them in letter and spirit.
- We inspire others to commit themselves to the principles of Responsible Care".

In 2007, the CCPA again reviewed its ongoing initiative and concluded that public concern about sustainability principles had increased and there was now an expectation that the private sector would reduce its impact on the environment as it developed economically [14, 27]. Taking a proactive posture, the CCPA looked to improving people's lives and the environment while striving to do no harm. The result of this extensive consultation between the CCPA, Board members, the advisory panel and industry leaders was a restatement of the RC commitment as follows [28]:

The Responsible Care® Ethic & Principles for Sustainability state that: "We are committed to do the right thing, and be seen to do the right thing. We dedicate ourselves, our technology and our business practices to *sustainability, the betterment of society, the environment and the economy*. The principles of Responsible Care are key to our business success, and compel us to:

- Work for the improvement of people's lives and the environment, while striving to do no harm.
- Be accountable and responsive to the public, especially our local communities, who have the right to understand the risks and benefits of what we do.
- Take preventative action to protect health and the environment.
- Innovate for safer products and processes that conserve resources and provide enhanced value.

- Engage with our business partners to ensure the stewardship and security of our products, services and raw materials throughout their life cycles.
- Understand and meet expectations for social responsibility.
- Work with all stakeholders for public policy and standards that enhance sustainability, act to advance legal requirements and meet or exceed their letter and spirit.
- Promote awareness of Responsible Care®, and inspire others to commit to these principles."

In addition, the CCPA added the following to its logo: "Responsible Care: Our commitment to sustainability."

1.4.2 Integrating the social dimension

RC's goal of developing public trust necessitated a focus on addressing environmental and health concerns. In 1983 the United Nations appointed an international commission to examine the state of the environment globally and to propose strategies for improvement. This commission, chaired by Norwegian Prime Minister Gro Harlem Brundtland, culminated in the publication of the report "Our Common Future" [29] focused on the interrelationships between the economy and the environment. The report underlined the crucial role played by the social dimension in achieving this precarious balance.

The social dimension was, therefore, never far from the surface of RC thinking. While a causal relationship between the chemical industry and its environmental impacts could be drawn, the social dimension and its effects on society presented a challenge to the entire industry and possibly all agents of economic society, in the larger context of corporate social responsibility. In cooperation with a number of other Canadian industry associations, the CCPA concluded in 1994 that social responsibility had to be examined on two levels: (1) a societal level where all stakeholders must reach consensus on their ever-evolving needs and the allocation of resources to each stakeholder in meeting them; this can include consideration of poverty, third world development, human rights, and many others and (2) a sectoral or company perspective where industry must preserve and grow its natural and human capital.

In 1994, the CCPA published "A primer on Responsible Care and Sustainable Development" [30] in which it tried to describe both the cohesion of these two concepts and areas where the scope of sustainable development extended beyond RC. A review of sustainable development principles was conducted while integrating organizations such as the International Chamber of Commerce Business Charter for Sustainable Development, the Canadian National Round Table on the Environment and the Economy and the Economy Objectives for Sustainable Development, Agenda 21, of the United Nations Conference on Environment and Development [30]. It chose 18 sectoral themes common to many of these organizations and, in its primer provided an attempt to describe visually their relationship with RC. There is a high degree of congruence between the themes of sustainable development and those of RC as related to the economic – environmental interface. However, certain social elements are truly societal in nature and beyond the scope of direct company actions.

As elements of social concerns were identified by the public through the CAPs, they were integrated into each company's responsibility portfolio. Community concerns have varied greatly from company to company, because of their operations, resulting in a diversity of social integration by individual companies. While RC had initially focused on Environmental Health and Safety (EH & S) concerns, it was often the more social issues at the local plant level that had to be addressed first.

1.4.3 Transition at the international level: Global expansion

In 1988 the CCPA presented its RC concept to representatives of the US Chemical Manufacturers Association (now the American Chemistry Council) and it was adopted by that association. Soon after, an effort was initiated to bring together all national chemical associations, with the intention of establishing RC worldwide. This led to the formation of the International Council of Chemical Associations (ICCA) and the formation of the ICCA Responsible Care Leadership Group, which exercises guardianship over RC globally and connects with other international institutions such as the United Nations Environment Program (UNEP) [1].

In 2004, a study, including an external stakeholder survey [31] conducted by the ICCA into RC development and implementation at the global level, led to the adoption of a set of Global Responsible Care Core Principles within a Responsible Care Global Charter as follows [1]:

"The Global Responsible Care Core principles commit companies and national associations to work together to:
- Continuously improve the environmental, health and safety knowledge and performance of our technologies, processes and products over their life cycles so as to avoid harm to people and the environment.
- Use resources efficiently and minimize waste.
- Report openly on performance, achievements and shortcomings.
- Listen, engage and work with people to understand and address their concerns and expectations.
- Cooperate with governments and organizations in the development and implementation of effective regulations and standards, and to meet or go beyond them.
- Provide help and advice to foster the responsible management of chemicals by all those who manage and use them along the product chain."

"Additionally, every national Responsible Care program must include eight fundamental features:
- A formal commitment by each company to a set of guiding principles, signed, in most cases, by the Chief Executive Officer.
- A series of codes, guidance notes and checklists to help companies fulfill their commitment.
- The development of indicators against which improvements in performance can be measured.
- Open communication on health, safety and environmental matters with interested parties, both inside and outside the industry.
- Opportunities for companies to share views and exchange experiences on implementing Responsible Care.
- Consideration of how best to encourage all member companies to commit themselves to, and participate in, Responsible Care.
- A title and logo which clearly identify national programs as being consistent with, and part of, the Responsible Care concept.
- Procedures for verifying that member companies have implemented the measurable or practical elements of Responsible Care."

In June 2008 [32], the ICCA Leadership Group unanimously adopted the concept of changing the official global identity of RC "Responsible Care for Sustainability."

1.5 Implementation challenges in gaining company support

In examining the development of RC in Canada, it is important to define what drove industry CEOs to pursue this approach and the challenges overcome during initial implementation.

1.5.1 Reflection on drivers

The factors that drove the creation of RC by the Canadian chemical industry are multiple. At the outset, the initiative came from the need to enhance public trust in order to maintain the industry's license to operate. Over time the industry's CEOs came to appreciate the intrinsic value of moving towards responsible behavior. From this base, a first set of interlaced stakeholder drivers were focused upon.

The Canadian chemical industry strongly believed that RC was an ethical and cultural approach to the issues and CEO/Board commitment became the cornerstone of success and the driver within individual companies. This helped to cement company-wide commitment developed through implementation activities, review processes and job performance criteria. Employee demand surfaced as a driver when pride in their company's commitment to RC improved employee morale, performance and recruitment [11, 33].

Building trust with environmental groups and NGO/Advocacy Groups through the the National Advisory Panel is also an important driver, which has been transparent from the outset and maintained the NAP's independent oversight [34].

Because the CCPA is mainly focused on manufacturers of industrial chemicals, member companies, rather than the public, were the consumers of the industry. The product stewardship aspect of RC, or supply chain management, became an adaptation to public consumer/customer demand and the means whereby companies could identify other member's commitment to the RC ethic. This concept is represented in the verification process which ensures that non-compliance by some members does not create problems for all [35].

1.5.2 Securing initial commitment

The principle that guided the initial commitment to RC was "evolutionary buy-in". All of the actions required by RC could not be defined at the outset without proper preparation of CEO's. It was necessary, therefore, to allow participants to gain confidence with each step and an understanding for the necessity of taking the next.

Securing and maintaining commitment has been an ongoing process. Several crucial actions were taken at the outset, which allowed the process of adoption of RC to move forward.

The CEO's of larger companies identified the threat to their continuing mandate to produce and worried about the threat of an inflexible regulatory regime that did not encourage progress beyond the regulations and stifled industry evolution [11]. They came to the conclusion that the economic bottom line could no longer be kept distinct from its environmental responsibility. In turn, they took on the leadership role and responsibility of convincing their peers. Eventually, in 1984 they helped develop the "Statement of Guiding Principles of Responsible Care" The members of the CCPA signed on to the statement and agreed to seek the voluntary commitment of all other CEO members of the CCPA. This work was completed in early January 1984, pre-dating Bhopal by 11 months.

In RC's early years, public opinion of the industry was negative and much of the public believed that the industry knew of problems inherent in its products, but hid this information from the public [36]. They believed that the industry dealt responsibly with its products during the manufacturing stages but concern increased further into their life cycle, during transportation, distribution, use and final disposal. Seminars with CEO's and other company employees were held to increase the process of sensitization. In an unprecedented step, the CCPA Board Chair met with editorial Boards of major Canadian newspapers and provided them with the highly critical study results. By presenting the results in a factual, if unflattering, manner, the industry took a crucial first step in establishing credibility with the public [23].

1.5.2.1 The need for risk communication

The importance of appropriate risk communication was emphasized from the inception of RC. Conventional wisdom within the industry argued that public misunderstanding of the industry was a result of the public's inability to comprehend complex science. What the leaders of industry failed to recognize was that public trust had to be earned. The CCPA consulted with experts in the field of risk communication. One was Professor Peter Sandman, formerly of Rutgers University, who developed the equation "Risk = Hazard + Outrage" [37]. Another was Professor Vincent Covello of Columbia University, who instructed industry leaders on conducting public consultation, with a focus on building trust through empathetic listening. These workshops were instrumental in convincing CEO's of the impact of public opinion and how to influence it [10].

1.5.2.2 The business case for Responsible Care

Early in the process of adoption of RC principles by the Canadian chemical industry some smaller companies expressed concern that RC could add to their costs and perhaps render them uncompetitive. The leader CEOs reinforced that RC was not idealistic, but rather a commitment to ethical practice. Leader CEOs offered help in designing systems and even tutoring to member companies, demonstrating the business case for RC and corporate social responsibility [33].

Leaders in the industry began to recognize a new array of opportunities through their responsible efforts, which became economic drivers. The focus on sustainable development issues meant companies could seize new markets through innovation and gain operational efficiencies and competitive advantages. As a result, this new set of potential advantages became a second wave of drivers for the leading companies in the chemical industry [11, 38].

1.5.2.3 Corporate social responsibility

Finally, it is important to examine social responsibility and its position as a driver. The recognition of social responsibility in capital markets has been a niche element that is still in its gestation period. As a result, many companies have not yet seen this as a driver for stock performance and for access to capital for growth. Demand for Socially Responsible Investing (SRI) is still in its infancy, but steadily growing. There is a strong belief that this will be a successful investment strategy in the near future and will become a key driver [39]. At this time, only a few companies are seeing the full potential of this opportunity in providing value for their shareholders. This is evidenced in the various SRI indexes including the Dow Jones Sustainability Index.

These drivers represent a generalization for a broad sector of industry. Various companies within the sector evolved at different rates and responded to the array of drivers with different intensities and timing.

1.5.3 Operational challenges

RC's focus on ethical concepts, as opposed to the more traditional regulatory compliance approach, has meant that the mindset of many member companies had to be radically challenged and modified during the initial implementation of the concept.

1.5.3.1 Overcoming a legal mentality

Initial attempts at writing guidelines governing chemical industry practice had been rejected on the advice of legal counsel who advised that subscribing to such guidelines would create undue liabilities. However, the context was such that regulatory pressures were building to a point where CEOs began to worry that the resulting constraints on their operations would essentially withdraw their mandate to produce. The CEOs came to realize that the choice was not between regulations or their absence, but rather among approaches designed to achieve certain defined objectives. If the objectives could be met strictly by industry's own efforts and outcomes were demonstrable, visible and measurable, the resultant market would be cost-efficient and flexible in its management. If the process was to be initiated before the implementation of regulatory constraints, there may be further positive benefits in terms of public recognition. Given that the legal and regulatory context is not static, meeting today's legal commitments did not mean one would meet later demands and remain competitive [3, 11].

1.5.3.2 Ethical behavior has a societal context, upon which trust will be built through external acceptance

Some companies believe that their singular responsibility is to create economic wealth. This can re-enforce an "us-them" mentality in the rest of society and the concern that industry cannot be trusted to respect greater social interests [38]. In contrast, the chemical industry set out to enhance recognition of the companies' wider social responsibilities to position itself as a part of the solution rather than the problem, by balancing economic objectives with social responsibility.

Behind this was the recognition that companies had to accept the concept that "Perception is Reality". Professor Peter Sandman's equation of "Risk = Hazard + Outrage" [37] states that hazard reflects scientific aspects of operation, and outrage combines social factors into the equation. Outrage increases when individuals feel controlled by others, or feel that a situation is unfair and morally relevant, or when sources are considered untrustworthy and when processes are considered unresponsive. According to this analysis, the chemical industry must listen, understand, and address public concerns. It must be transparent and responsive to community inquiry in order to gain public trust.

1.5.3.3 Seeking and acquiring full Chief Executive Officer commitment

The initial reaction to RC in the chemical industry was mixed. Some saw it as just another initiative in a series of environmental management programs to be implemented over the years. Responsibility for its implementation could be assigned to an employee, and interest in it would be time-limited. Alternatively, some saw it as new regulations imposed by the industry association [4]. These misapprehensions needed to be addressed if the initiative were to bring about a lasting change in company and industry behavior. RC had to be embraced by CEO's, who in turn had to signal clearly to employees that RC would form their companies' guiding principles going forward. One method was the implementation of performance evaluations [34, 40] based on the new mentality. The CCPA recognized that changing the culture of the industry required time for absorption so that the strength of its success would come from employees themselves.

Another assumption that had to be overturned was the desire for a prescriptive implementation plan with fixed criteria and endpoints. Instead, RC requires that member companies consult with stakeholders as to what "the right thing to do" would be in their circumstance, and to do so in a publicly accessible fashion. Individual CEOs must affirm that his or her company has taken all practical precaution to ensure its products do not present an unacceptable level of risk to its employees, customers, the public, or the environment. Meeting the codes of RC is a constant, dynamic process without endpoints and becomes the culture of operation rather than a prescribed program [17, 40].

1.5.3.4 Acceptance of the need for comprehensiveness

Initially there were concerns about the resources required to adhere to RC codes in all areas of operation. However, since it is cultural and not programmatic in nature, RC must pervade all areas of operation. An important part of the evaluation is the measure of RC "embeddness" in the company structure and corporate culture. One of the ways evaluators determine how well an RC company has incorporated the ethic into their mandate is to measure the employee buy-in that exists in the company. This allows the evaluator to gain a better understanding as to how well RC is being implemented in a bottom-up fashion [41].

1.5.3.5 Recognizing that the weakest link could defeat the initiative

Each company has a stake in each others' performance, necessitating cooperation and mutual assistance in implementation of RC codes. Expertise was shared through workshops and one-on-one help between CEO's. This mutual help concept is what fostered small company adherence to the initiative. For example, RC proved to be very successful for Sulco Chemicals, the small company test-pilot for the implementation of RC. This helped increase the adoption of RC in smaller organizations across Canada [42].

RC codes are a way of demonstrating the care with which companies handle their products. For companies who meet the codes, this may be used in a court of law in a defense of due diligence demonstrating that they have been following accepted industry practice as reviewed by external stakeholders. Those who do not adhere to the codes may leave themselves vulnerable to legal prosecution [33].

1.5.3.6 Overcoming fears of Responsible Care as undue competitive burden

The concern voiced most often by company CEO's was that RC and other environmental management systems would cost too much money to implement [11, 16]. However, RC describes what a diligent company must do to ensure that its products do not present an unacceptable level of risk to its employees, customers, the public, or the environment. Its codes were designed by industry insiders, and the means of a company's compliance to the code is at the discretion of the individual company. Alternative, government-imposed regulations with the attendant negotiations and uncertainties are generally believed to be more costly [11].

Companies report that after reviewing their RC commitment and management systems with their insurers, rates were reduced. Some companies have reported bankers, worried about environmental liabilities, have responded positively to their RC commitment, resulting in lower financing rates. There have also been examples of expedited permitting, marketing advantages and lower workers' compensation premiums, and costs as a result of RC participation [33].

1.6 Maintaining commitment to Responsible Care

Having established the crucial role of public opinion and consultation during the initial phase of RC implementation, attention then turned to the best means by which the industry could address public concerns.

1.6.1 Guiding principles

This set of principles describes the commitment of the CCPA to RC. Initial requests to chemical industry CEO's for voluntary adoption of these principles were met with 95% compliance. Subsequently, subscription and adherence to these principles has been made a mandatory component of CCPA membership. Companies are also encouraged to prominently display this signed document within their organization. The commitment to guiding principles governs chemical operations; however, diversified companies are encouraged to implement the principles in their non-chemical operations as well. Today, non-chemical producers, such as chemical transporters and producers are welcomed as Responsible Care Partners with the same compliance requirements as regular members.

1.6.2 Codes of management practices

The codes describe in concrete terms the implementation of management practices which ensure products and associated processes and operations are handled safely throughout their life cycle. The codes were developed over three years by a team of functional experts from a cross-section of Canadian chemical companies. It was recognized that each member company would encounter unique challenges in meeting the codes' requirements, and therefore a rigid schedule of compliance was decided against. The Board of Directors of the CCPA advised member companies that they were expected to complete the process within three years of initiation in order to maintain public credibility. Every six months, companies report to the CCPA regarding compliance rates. As a final aspect, the six codes cover: Research and development, manufacturing, transportation, distribution, hazardous waste management, and community awareness and emergency response.

Conformance includes detailing how operations are carried out by associated groups such as suppliers, customers, transporters, or university labs conducting contract research for a company. Following completion of this procedure, the CEO must attest to the CCPA that code requirements have been met [17].

1.6.3 Advisory process

The CCPA secured public input into the RC development process by having a third-party consultant assemble a cross-sectional group of activists, academics, consumers, seniors, and youth to follow the development of RC from its inception. The independence of this consultative body and its inclusion of members critical of the Canadian chemical industry were vital to ensure the credibility of the panel. The National Advisory Panel continues to meet today with rotating personnel. Its advice has been sought before every important decision being taken by the Board of Directors of the CCPA. For example, each code of practice was submitted to the panel before going to the Board for final approval. Additionally, an unedited message from the National Advisory Panel is included in the annual report of the association on RC. Beyond the national level, every member company site must have a Community Advisory Process in place to meet the Community Awareness and Emergency Response code [17].

1.6.4 Chief Executive Officer forum

RC calls for a new mentality embraced by member CEO's and entrenched within each company. The CEO must back his or her commitment by ensuring that performance of individuals will be evaluated along the lines of this new mentality. CEOs are also best placed to convince their colleagues of the merits of compliance with RC. These industry leaders are subdivided into groups of 8 to 15 by region and meet at least four times a year to monitor the compliance progress of each member and exert peer pressure to ensure that efforts are constant.

1.6.5 Results confirmation process

To counter the argument that RC is simply a public relations tactic, CCPA initiated two important activities. First is third party verification by a four-member team, comprising:

–   Two experts in the implementation of RC. These are often people who have retired from the industry after being coordinators themselves.
–   One expert in environmental matters. Often these have been professors of environmental science or environmentalists familiar with the development of RC.
–   One representative of the community in which a company operates. This can be anyone with credibility in the community, for example a high school principal, or activist. The role of this individual is to confirm to the community that the process has indeed been a serious one.

The second activity in the results confirmation process is the production by each participant company of an annual report of all air, water, and ground emissions, including waste and a five-year forecast. This Reducing Emissions Report is mandatory for all companies and results must be made available to community stakeholders [43]. The CCPA also provides information to the public on its overall progress in RC including the unedited message of the National Advisory Panel.

1.6.6 Name protection

The name "Responsible Care" and its logo have become recognized as one of the most significant assets of the chemical industry, and in recognition of this, are copyright protected. Only companies who are willing to participate in adoption of RC codes and culture are permitted to use these symbols [3].

1.7 Role of government

While RC is an initiative conceived and implemented by industry, government has indicated it considers RC a valid, credible alternative to moderate regulatory pressures. In 2001, a memorandum of understanding was signed by the Ministers of Industry and Environment and the President of the CCPA to meet periodically at the senior officials level, to review progress by the industry in its efforts at emission reduction, and to discuss RC in general [44]. The Environment and Industry departments have encouraged other economic sectors to follow the chemical industry's example and implement initiatives similar to RC. In 1990, Environment Canada nominated RC to receive the United Nations Environmental Program's Global 500 Roll of Honour Award for outstanding contribution to environmental improvement. RC was the first industry sector to receive the award, for what UNEP described as "Outstanding practical achievements in the protection and improvement of the environment" [45].

1.8 The future of Responsible Care in Canada

The American Chemistry Council (ACC) has acknowledged the shortcomings the American Responsible Care system, which differs significantly from the Canadian implementation, specifically in that it did not have a verifiable set of standards to evaluate the performance of firms and their adherence to RC. Many chemical associations, such as the CCPA, have recognized that insufficient third party monitoring of a firm's performance has been a major deterrent to the success of RC. In response to these issues, as of 2007, the ACC requires firms to subscribe to a hybrid Responsible Care/ISO 14001 environmental management system or a revamped Responsible Care Management System [6, 46]. The ACC's RC-14001 overlays RC codes onto ISO 14001 codes and requires that only one evaluation be done for both standards simultaneously. This type of policy is a major step towards legitimizing RC and ensuring transparent reporting and true accountability in the United States [4, 46].

In Canada, verification is already an integral part of the process; however, checklist-based monitoring has not been the direction that the initiative has moved towards, largely because it fails to measure the extent to which the ethic has been integrated into the corporate culture of a firm [47]. Given the Canadian RC's movement towards the integration of precautionary policies and the use of inherently safer technologies, one possible path for the evolution of the ethic is to require companies to report on standardized indicators such as those outlined in the G3 Sustainability Indicator Guidelines of the Global Reporting Initiative (GRI), which would complement the reporting already required through NERM. Rather than hybridize the verification process, such as was done in the US, the CCPA could explore hybridizing the GRI with NERM, which could establish a more robust reporting standard.

The United Nations developed the GRI in order to ensure that companies which enroll in the program report on all relevant indicators and not just on the ones in which they show positive performance [48]. According to Tchopp [39], there are many examples of companies that report on their environmental successes but not on their failures. This can give the public an incomplete picture of the company's social responsibility strategy.

It is important to consider the issue of transparency and standards for companies who embark on environmental management programs and sustainable initiatives. Reporting standards provide investors and consumers with a transparent basis of comparison for various companies. At the same time companies can use standards to gauge their success and properly target necessary improvements which could lead to even greater efficiencies and profitability [48].

Reporting standards can also help identify the degree to which a company's sustainable initiatives are holistic and integrated. Positive performance on most indicators would indicate integrated strategies as opposed to narrowly focused ones. An integrated strategy is proven to yield greater profits and would be a more popular choice for a socially responsible investor [49]. This type of measurement would align better with the Canadian RC's goal to have the ethic embedded in the corporate culture rather than be more prescriptive and checklist-based, as is the case with other environmental management systems.

1.9 Conclusion

RC is a dynamic statement of ethical concern, in a continual state of evolution in Canada. It was initiated in the early part of the 1980s as a simple one-page statement of principles touching the singular aspect of environmental responsibility. It has since evolved into a set of codes, verification processes, visible performance measurement, and a deeper understanding of the principles first stated. There has been one constant; building trust through ethical behavior, listening attentively to the evolving concerns of the public and providing responses that clearly demonstrate concerns have been taken seriously.

Sustainable development principles in the RC ethic have become integrated in decision-making processes. The Canadian Chemical Producers' Association, by pursuing a dynamic and evolving RC, enacts its ongoing commitment to meeting the sustainable development aspirations of Canadian society. In 2007 the CCPA Board reviewed the state of RC in Canada and determined that they were largely invisible on the issues of sustainability and global warming; issues of major concern to the Canadian public. Once again, the leaders on the Board stepped forward and ensured that these issues would be integrated explicitly within the ethic of RC in the future.

It has taken leaders amongst industry CEOs on the RC Board, frontline workers who implement the ethic in their daily work, and citizens who have provided input and oversight to push the borders of conventional thinking and convince other companies to commit to do the right thing. These leaders realized that the health of the chemical industry was at stake and responded by establishing a transparent, open, and responsive chemical industry whose commitment to public safety and consultation is well documented and genuine.

Acknowledgements

This research was made possible by the International Union of Pure and Applied Chemists. We particularly thank Bernard West and Brian Wastle for their review and comments on earlier drafts of this chapter.

References

[1] Responsible Care Global Charter. International Council of Chemical Associations, 2006. (Accessed March 29, 2009, at http://www.icca-chem.org/Global/Initiatives/RC_GlobalCharter2006%5B1%5D.pdf?epslanguage#equal#en.).

[2] West, B. Responsible Application of Chemistry: An Introduction to Responsible Care. International Union of Pure and Applied Chemists, 2007. (Accessed March 29, 2009, at http://old.iupac.org/projects/2006/2006-047-1-022.html.).

[3] Responsible care in your neighborhood. Canadian Chemical Producer's Association, Ottawa, 2005. (Accessed March 29, 2009, at http://www.ccpa.ca/files/Library/Reports/RCoverviewEN.pdf.).

[4] Two decades of responsible care: Credible response or comfort blanket? Environmental Data Services (ENDS) Ltd., No. 360, London, 2005. (Accessed March 25, 2007, at http://www.responsiblecare.org/filebank/ENDSFeaturesResp_Care.pdf.).

[5] Bélanger, J. Responsible care in Canada: The Evolution of an Ethic and a Commitment. Chemistry International 2005;27(2):1.

[6] Schmitt, B. Industry's critics make respect an elusive goal. (Responsible Care). Chemical Week 2002;164(27):42(3)–5.

[7] Moffet, J., Bregha, F., Middelkoop, M. J. Responsible care: A case study of a voluntary environmental initiative. In: Webb, K., ed. Voluntary codes: Private governance, the public interest and innovation. Carleton Research Unit for Innovation, Science & Environment, Ottawa 2004:177–208.

[8] Dudok van Heel, O. Buried treasure: Uncovering the Business Case for Corporate Sustainability. SustainAbility, London, UK, 2001.

[9] Heath, R. L., Bradshaw, J., Lee, J. Community Relationship Building: Local Leadership in the Risk Communication Infrastructure. Journal of Public Relations Research 2002;14(4):317–53.

[10] Covello, S. T. Risk Communication. In: Frumpkin, H., ed. Environmental Health: From Global to Local. John Wiley and Sons, New York, 2005:988–1009.

[11] Willard, B. The Next Sustainability Wave: Building Boardroom Buy-in. New Society Publishers, Gabriola, BC, 2005.

[12] Reisch, M. Responsible care: The chemical industry has changed its ways, but more radical actions may be needed to change public opinion. CENEAR 2000;78(36):21–6. (Accessed May 2008, at http://pubs.acs.org/cen/coverstory/7836/7836bus1.html.).

[13] Peters, R. G., Covello, V. T., McCallum, D. B. The Determinants of Trust and Credibility in Environmental Risk Communication: An Empirical Study. Risk Analysis 1997;17(1):43–54.

[14] News Release – CCPA Launches Important Responsible Care Ethic. Canadian Chemical Producer's Association, 2008. (Accessed March 29, 2009, at http://ccpa.ca/News/news10200801.aspx.).

[15] McDonough, W., Braungart, M., Anastas, P. T., Zimmerman, J. B. Applying the principles of green engineering to cradle-to-cradle design. Environmental Science Technology 2003;37(23):A434.

[16] Hawken, P., Lovins, A., Lovins, L. H. Natural capitalism. Little, Brown & Company, New York, 1999.

[17] The Ethic and Codes of Practice of Responsible Care. Canadian Chemical Producer's Association, Ottawa, 2000.

[18] Canada's Invisible Industry. Canadian Chemical Producer's Association, Ottawa, 1973.

[19] CCPA Report on Public Understanding of the Industry. Canadian Chemical Producer's Association, Ottawa, 1979.

[20] Allen, D. K., Havey, M. C., Dickie, A. Derailment: the Mississauga Miracle. Government of Ontario 1980.

[21] Laughton, R. V., Kierstead, T., Moran, T. S., Munro, S. Canadian Environmental Programs in the Petroleum and Petrochemical Industry: Programs, Policies and Priorities. Pollutech Group, 1994. (Accessed March 29, 2009, at http://www.pollutechgroup.com/papers/p8.htm.).

[22] Guiding Principles to be Developed. Canadian Chemical Producer's Association, Ottawa, 1978.

[23] Bélanger, J. Personal observation.

[24] Emissions Reporting (NERM) Toolbox. Canadian Chemical Producer's Association, 2009. (Accessed March 29, 2009, at http://ccpa.ca/Issues/Enviro/NERMtoolbox.asp.).

[25] Statement of Commitment to Responsible Care. Canadian Chemical Producer's Association, Ottawa, 1984.

[26] Revised Statement of Commitment. Canadian Chemical Producer's Association, Ottawa, 2003.

[27] Panel challenges industry to make new principles a reality. Canadian Chemical Producer's Association National Advisory Panel, 2009. (Accessed March 29, 2009, at http://ccpa.ca/files/Library/Documents/RC/PanelLetter2009Final.doc.).

[28] Responsible Care. Canadian Chemical Producer's Association, 2008. (Accessed March 29, 2009, at http://www.ccpa.ca/ResponsibleCare/.).

[29] World Commission on Environment and Development: Our Common Future, Brundtland Commission, Oxford University Press, Oxford, 1987.

[30] A primer on Responsible Care and Sustainable Development. Canadian Chemical Producer's Association, Ottawa, 1994.

[31] External Stakeholder Survey: Final Report for the Global Strategic Review of Responsible Care. SustainAbility, Washington, DC, 2004. (Accessed May 2008, at http://www.responsiblecare.org/filebank/SUSTAINABILITY%20Stakeholder%20Survey%20Final%20Report.pdf.).

[32] Responsible Care. International Council of Chemical Associations, 2009. (Accessed March 29, 2009, at www.responsiblecare.org.).

[33] Does Responsible Care Pay? Canadian Chemical Producer's Association, Ottawa, 2005.

[34] Message for 2002/2003 responsible care report. Canadian Chemical Producer's Association National Advisory Panel, Ottawa, 2003. Ottawa. (Accessed March 29, 2009, at http://ccpa.ca/files/Library/Documents/RC/CCPA_Panel_message_2003.doc.).

[35] Verification. Canadian Chemical Producer's Association, 2008. (Accessed March 29, 2009, at http://ccpa.ca/ResponsibleCare/verification.asp.).

[36] King, A. A., Lenox, M. J. Industry self-regulation without sanctions: The chemical industry's responsible care program. Academy of Management Journal 2000;43(4):698–716.

[37] Sandman, P. M. Responding to Community Outrage: Strategies for Effective Risk Communication. Fairfax, VA: American Industrial Hygiene Association, 1993.

[38] Elkington, J. The Chrysalis Economy: How Citizen CEOs and Corporations can Fuse Values and Value Creation. Capstone Publishing Ltd., Oxford, UK, 2001.

[39] Tschopp, D. J. Corporate social responsibility: A Comparison between the United States and the European Union. Corporate Social Responsibility and Environment Management 2005;12:55–9.

[40] Responsible Care Re-verification 2005–2008. Canadian Chemical Producer's Association, Ottawa, 2005.

[41] Reisch, M. Taking the next step: The US responsible care program takes on security and verification issues as keys to enhancing its longevity and reputation. CENEAR 2003;81(21):15.
[42] Chartrand, H. F. Three Case Studies on the Impact of Responsible Care. Canadian Chemical Producer's Association, Ottawa, 2005.
[43] Reducing Emissions Report. Canadian Chemical Producer's Association, 2009. (Accessed March 29, 2009, at http://ccpa.ca/files/Library/Documents/Environment/Reducing_Emissions_16_FINAL.pdf.).
[44] Memorandum of Understanding (MOU) with the Canadian Chemical Producers' Association, Environment Canada, 2001. (Accessed March 29, 2009, at http://www.ec.gc.ca/epa-epe/ccpa-acfpc/en/details.cfm.).
[45] Global 500 Forum: Roll of Honour/Laureates. United Nations Environment Program, 1990. (Accessed March 29, 2009, at http://www.global500.org/ViewLaureate.asp?ID#equal#410.)
[46] AmericanChemistry.com: Responsible care®. American Chemistry Council, 2007. (Accessed July 9, 2007 at http://www.americanchemistry.com/s_responsiblecare/sec.asp?CID#equal#12 98&DID#equal#48.
[47] Wartle, B. Personal communication. May 27, 2008 (details available on request from the authors).
[48] Global Reporting Initiative. G3 Sustainability Reporting Guidelines, 2006.
[49] Perrini, F., Tencati, A. Sustainability and stakeholder management: The need for new corporate performance evaluation and reporting systems. Business Strategy and the Environment 2006;(15):296–308.

Peter Topalovic, Gail Krantzberg, and Maria Topalovic

2 Responsible Care in global supply chains: A case study

Abstract: This chapter provides an analysis of a case study in which 80 Haitian children died of a poisoned cough syrup tainted with diethylene glycol. Issues with the management of a supply chain that originated in China, continued through Germany and the Netherlands and terminated in Haiti ultimately led to the avoidable tragedy. The analysis addresses the possible roles of the Chemical Industry's corporate governance initiative, Responsible Care, in the supply chain and asks the question: Would Responsible Care, appropriately applied throughout the supply chain, have averted the crisis experienced in Haiti and other developing countries? Stakeholder trust, corporate due diligence and corporate culture are the key topics examined in the context of three main theories which lead to the conclusion that Responsible Care would have averted the crisis. The three main theories investigated are: (a) The companies involved did not internalize the concepts of product stewardship and the cradle to cradle philosophy that Responsible Care advocates; (b) the loss of business and reputation through erosion of trust was not a major consideration in the decision-making processes of the companies involved; and (c) the major players involved in the case study did not embed the philosophies of the Responsible Care ethic or create a corporate culture of protection of all stakeholders.

2.1 Introduction

The issues surrounding global supply chains which originate and terminate in the developing world continue to plague the chemical industry since first making news in the United States in 1937 and Haiti in 1996, when contaminated cough syrup killed 85 children; they continue to make news into the new millennium in countries such as Panama and Nigeria [1, 2]. In North America, issues with lead poisoned toys, diethylene glycol (DEG)-tainted toothpaste and other dangerous products mimic the problems encountered in developing countries; however, in most situations, tainted products are identified before they reach the end customer. Tough North American regulations and industry standards which are not usually present in developing countries may be responsible for successfully identifying the dangerous products before they pose a threat [3, 4]. This case study examines the effects of DEG poisoning and drug counterfeiting which occurred in the 1996 Haitian case. A mix of national and international regulations, along with corporate voluntary initiatives is assessed as a way to determine whether the public can be better protected from these poisonings and tragedies.

The emphasis of the study will be on the International Chemical Industry's Responsible Care (RC) ethic, an environmental management system which compliments regulations and drives a company's performance beyond the minimal standard. It was developed in the 1980s to meet the threat of costly new regulations and decreasing public trust in the chemical industry. It consists of an ethical set of principles as well as management system of codes and practices which aim to address environmental, ethical and social concerns of chemical manufacture over the entire life cycle [5, 6]. Throughout this paper, the use of the term RC will refer to the Canadian implementation as outlined by the Chemistry Industry Association of Canada (CIAC), formerly the Canadian Chemical Producers Association (CCPA), in the "Ethic and Codes of Practice of Responsible Care" [7]. This includes elements such as external third party verification of compliance and the combination of ethical and prescriptive measures in the implementation of the code. The central question of this case study asks: *Would Responsible Care, appropriately applied throughout the supply chain, have averted the crisis experienced in Haiti and other developing countries?* To answer this question and determine whether RC can have a positive impact, an investigation of global supply chains, due diligence, stakeholder trust and corporate culture follows. This case study analysis will examine evidence to determine whether or not the following theories are correct:

- The companies involved did not internalize the concepts of product stewardship and the cradle-to-cradle philosophy that RC advocate.
- The loss of business and reputation through erosion of trust was not a major consideration in the decision-making processes of the companies involved.
- The major players involved in the case study did not embed the philosophies of the RC ethic or create a corporate culture of protection for all stakeholders.

2.2 Industrial chemical supply chains and product stewardship

The supply chain is a term which defines the development and use of a product from materials extraction, manufacture, distribution, use, end of life and reuse of a product or material. The chain can be broadly segmented into two main processes:

- Production, planning and inventory control which involves materials extraction, manufacturing and storage.
- Distribution and logistics which involves scheduling, transport from storage to other distributers, retailers and eventually to the end user [8].

Fig. 2.1 describes this process.

The Environmental Protection Agency's (EPA's) lean and green supply chain involves another process to complement the traditional supply chain, which is referred to as reverse logistics. This includes life cycle analysis, product take back, and cradle-to-cradle philosophies rather than cradle-to-grave philosophies [9]. In Fig. 2.2, reverse logistic considerations are augmented to the typical supply chain outlined in the dashed bubble.

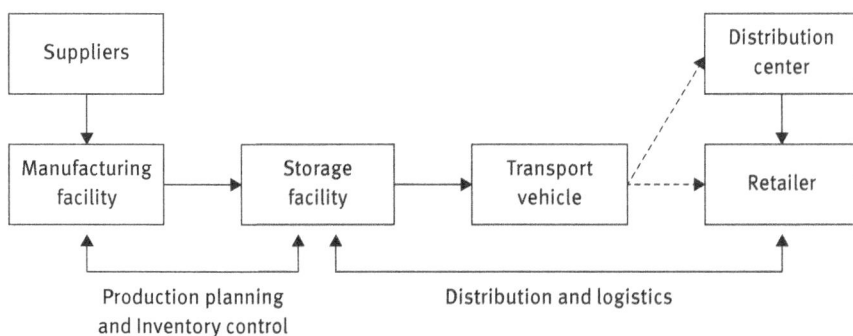

Fig. 2.1: The supply-chain process. Adapted with kind permission from Beamon, International Journal of Production Economics, Elsevier, 1998 [8]

According to the International Council of Chemical Associations (ICCA) [10], product stewardship involves the integration of health, safety, and environmental considerations into the life cycle of the product. McDonough et al. [11] extend this principle with the idea of cradle-to-cradle design. This involves processes which aim to eliminate waste and harm to stakeholders by turning a waste product into a valuable input to another process; thereby creating a financial incentive to reduce waste and ecosystem impacts.

Designs that incorporate product stewardship considerations are proactive, precautionary and more sustainable than designs that require end-of-pipe solutions to mitigate their negative effects [11]. Product stewardship is a highly integrated, multi-disciplinary subject. It is prescriptive and rule based, while also promoting a philosophy of preventative, responsible design. At its root are the concepts of due

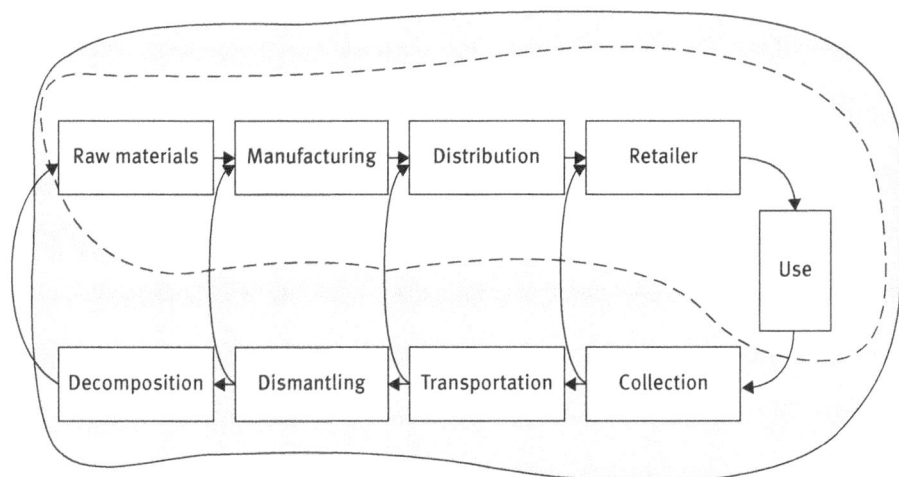

Fig. 2.2: Lean, green supply chain. Adapted with kind permission from Kainuma and Tawara, International Journal of Production Economics, Elsevier, 2006 [12]

diligence, corporate social responsibility, extended producer responsibility, product life cycle considerations and profitability. A company who internalizes this concept understands that downstream uses of their product are their responsibility regardless of how many times the product exchanges hands; this is the cornerstone of RC, according to the CIAC [5].

2.3 Overview of the case study

In 1996, 85 Haitian children died from ingesting cough syrup tainted with DEG. DEG, an industrial chemical found in anti-freeze and other products, is a colorless and odorless liquid, similar in appearance and viscosity to glycerin, a common base in many pharmaceutical preparations [1]. According to O'Brien et al. [13], the Acute Renal Failure Investigation Team, the group charged with analyzing the DEG poisoning outbreak, found that the median estimated toxic dose of DEG for a patient was 1.34 mL/kg (range, 0.22–4.42 mL/kg). There is no known minimum toxic dose, but doses of 1 mL/kg or more have proven to be lethal [13]. The mechanism for toxicity in humans is not known; however, 2-hydroxyethoxyacetic acid, the metabolic product of the enzyme aldehyde dehydrogenase, is considered to be a renal toxin [2]. The median concentration of DEG found in the tainted bottles of acetaminophen syrup was 14.4%. According to the United States Food and Drug Administration (FDA), the limit for DEG contamination is 0.1%, as outlined by the United States Pharmacopeia monograph [4].

An FDA investigation found that Pharval, a local company who produced the cough syrup products, Afebril and Valodon, did not contaminate the product at their site. Instead, a supposedly pharmaceutical grade shipment of glycerin, a key component in the most widely prescribed cough syrup in the country, was contaminated at its source in China. However, the Haitian company was under the assumption that the chemical was produced in Germany from chemical giant Helm AG, parent of VOS B.V., and as such, did not implement any quality controls on the imported European product [4, 14].

Issues surrounding product stewardship, supply-chain management, quality control, politics, and legal responsibility are at the heart of this disaster and combine together to create a complicated web of interactions leading to negative implications for all the parties involved. In the aftermath of this disaster, many of the companies in the supply chain engaged in finger pointing, lawsuits, and denial of responsibility.

2.3.1 Accountability for poor management of the supply chain and a lack of quality control

The United States FDA's investigation of the cough syrup incident indicated that the product batches were not tested for DEG contamination [15] and the pharmaceutical manufacturing and testing facilities at Pharval laboratories did not meet international standards [16]. This was due to a variety of factors, including the fact that Haitian regulations are not as strict as those in other countries. Furthermore, the maintenance costs for clean rooms, proper Heating, Ventilation and Air Conditioning (HVAC) systems and high-tech testing equipment are too high for most companies in developing countries [4].

The investigation of the glycerin suppliers leads to some disheartening conclusions related to the management of international supply chains in China, the Netherlands, and Germany, especially where developing countries are concerned. In the Haitian case, no company along the supply chain was found to be directly responsible for the problems that occurred. Some denied responsibility, others covered up mistakes and in the case of the contamination source it was never determined which Chinese company was responsible; however, it is believed to be Sinochem, based in Peking [3].

The investigation of VOS B.V. by the Special Rapporteur of the United Nations (UN) Economic and Social Council of the Commission on Human Rights [17] found that the company knew about the contaminated glycerin after it sent the shipment to Haiti, but did not alert authorities. VOS B.V. sent a sample of the Chinese shipment to an independent testing laboratory, and falsely marked the barrels of glycerin as 98 PCT USP, pharmaceutical grade.

A certificate of pharmaceutical quality also accompanied the barrels, which was taken from the certificate that originated from China [17]. It is common practice to re-use certificates when a product changes hands from manufacturer to supplier and onward to the destination customer [6, 18]. The identification of the source contamination and the FDA inspection of Pharval laboratories clearly established that the glycerin was contaminated at the source and not along the supply chain. It was a lack of product stewardship policies and a lack of due diligence across the supply chain that lead to the disaster. In Haiti, as well as in Panama and Nigeria, two more recent countries which experienced DEG contamination [1, 2], the factory's original certificate of analysis for the glycerin containers did not accompany them as they moved across the supply chain. Instead a copy of the original was used and stamped with the receiving company's information each time the container exchanged hands [6, 16]. An overview of the DEG contamination cases that occurred worldwide are summarized in Table 2.1.

Helm AG, one of the largest chemical companies in the world and the parent company of VOS B.V., declined to comment on the case, given that the contamination occurred outside of Germany. Helm AG has been associated with other issues involving the transport of materials to the third world, according to the German media [17]. The Chinese government also denied any responsibility since the glycerin was not shipped directly to Haiti from China.

While no international supply-chain management regulations exist to solve problems such as this one; the European Union's Registration, Evaluation, Authorization and Restriction of Chemical Substances (REACH) legislation and RC's supply-chain management policies may have a positive effect [19]. Furthermore, the World Health Organization (WHO) has developed good manufacturing guidelines for pharmaceuticals and material inputs [20]. This issue has received major coverage over the last ten years since the Haitian disaster occurred. In the Haitian case, there were many organizations responsible for transporting the glycerin across international boundaries and therefore, it is very difficult to lay blame. However, after the details of the case were sorted out, some litigation proceedings were undertaken in the Netherlands, Germany and Haiti.

2.3.2 Economic and social effects of the incident and the ensuing litigation

Pharval settled with the Haitian families whose children died from DEG exposure for $10 000 USD per family. They also filed a civil suit against VOS B.V. in the Netherlands, jointly with the families. The litigation against the companies involved focused on VOS B.V., since they knowingly sent DEG-contaminated glycerin to Haiti [17]. Eventually a civil suit was settled outside of court with the Dutch company for the same amount as Pharval had settled with the parents. In the aftermath, the affected families were compensated, yet no company accepted full responsibility for the tragedy. VOS B.V. was also prosecuted by the Dutch government, found guilty of the cover up and fined $250 000 USD [16].

Other attempted litigation has generally failed to produce a favorable outcome for the plaintiffs. The Chinese government and corporations would not work with the United States FDA, who assisted Haitian officials with the investigation, to find the source of the contamination, denying responsibility and moving glycerin operations away from the site of contamination. David Mishael, a lawyer in the United States who has spent the last ten years since the Haitian disaster representing Haitian parents, has unsuccessfully pursued legal claims against Helm AG and VOS B.V. [16].

In terms of the economic and social effects of the disaster and the ensuing litigation, the costs associated with non-compliance were $250 000 USD and very little transparency or accountability was demanded in the aftermath. Policies on supply-chain management remain largely unchanged world-wide, with the exception of the European Union's REACH legislation. Evidence of a policy and regulatory gap is clear in the past ten years of repeated DEG contamination in developing countries throughout the world. In many cases, the source of this contamination continues to be from poorly regulated Chinese suppliers.

Panama experienced a tragedy similar to that of Haiti in 2006, with DEG being the contaminant in a government manufactured cough syrup, resulting in hundreds of deaths [1]. Over the years, other countries such as Bangladesh, Argentina, Nigeria, China, and India have also been affected by DEG poisoning cases. The latest cases in

Table 2.1: Summary of Diethylene Glycol (DEG) Outbreaks [1, 2, 13]

Year	Country	Deaths	Route	DEG Vehicle	DEG Source
1937	United States	105	Oral	Sulfanilamide elixir	DEG excipient
1967	South Africa	7	Oral	Liquid sedatives	Unknown
1985	Spain	5	Topical	Sulfadiazine	DEG excipient
1986	India	14	Oral	Glycerin	Industrial-grade glycerin
1990	Nigeria	47	Oral	Acetaminophen	DEG replaced propylene glycol
1990–1992	Bangladesh	51	Oral	Acetaminophen	DEG replaced propylene glycol
1992	Argentina	26	Oral	Propolis	DEG excipient
1995–1996	Haiti	85	Oral	Acetaminophen	DEG replaced propylene glycol
2006	Panama	78	Oral	Acetaminophen	DEG replaced propylene glycol
2008–2009	Nigeria	54	Oral	Acetaminophen	DEG replaced propylene glycol

Panama and Nigeria [2] clearly illustrates that the proper steps have not been taken to minimize this preventable disaster. Table 2.1 summarizes these cases.

In response to Haitian experience, an inexpensive DEG testing kit was developed to assist regulators in identifying contaminated shipments; however, this is not in widespread use [21]. The test uses thin-layer chromatography (TLC) to detect and quantify DEG contamination, using visual inspection testers which cost $1.00 USD per test and do not require a laboratory environment. While the test is not accurate enough to detect 0.1% contamination, it would have avoided many of the cases presented in this analysis [2, 15].

DEG poisoning has now become a global problem with a preventable death toll in the hundreds. China has recently taken action against drug counterfeiters with a revised pharmaceutical law [22], but when the role of Chinese pharmaceutical companies in the disaster in Panama is examined, they continue to counterfeit and contaminate [1]. Many developing countries, including China and India need tougher regulations in order to meet safe standards [6].

After the Haitian incident, world health experts recommended improving the certificate of authenticity system to provide a clear path of the material flow through the supply chain from source to destination. It also stressed that transparency and accountability should be enforced through regulations and investigations within and between international borders. As long as counterfeiters do not fear prosecution, there is no incentive to improve the quality of their products [4, 13, 18].

2.4 Case study investigation: methods and analysis

This case study addresses questions that arise from this disaster in the context of RC. We test the hypothesis that the magnitude of this crisis would have been reduced if some or all of the companies along the supply chain subscribed to RC principles and codes. This will serve to evaluate the strengths and short comings of RC.

2.4.1 Theory 1: The companies involved did not internalize the concepts of product stewardship and the cradle-to-cradle philosophy that Responsible Care advocates

The RC ethic states:

> "We are committed to do the right thing and be seen to do the right thing. We are guided towards environmental, societal and economic sustainability … We are stewards of our products and services during their life cycles in order to protect people and the environment" [7, p.2]

The case study findings indicate that the absence of product stewardship considerations was a fundamental flaw in the supply chain that transported DEG-tainted glycerin to Haiti. To examine how RC would have avoided the issues of this unhealthy supply chain, each player's role in the chain is analyzed.

2.4.1.1 The Source: Sinochem in China

According to BUKO Pharma-Kampagne [3], a non-governmental organization that examines the activities of German pharmaceutical companies in the developing world, the most likely source company in the Haitian supply chain was Sinochem, in Peking, China. This company does not subscribe to RC, nor is it subject to regulations such as REACH. Sinochem manufactured the DEG-contaminated glycerin, most likely with the intention to create a cheaper, counterfeit product. This has been a common trend with drug preparations being exported from China and other developing countries and has grown over the last decade [20, 23].

In terms of RC, the company violated one of the guiding principles which states that companies shall "manufacture chemicals and chemical products in a manner which protects people and the environment from hazards" [7, p.22]. Furthermore, it would have issued fair and timely downstream warnings about the dangers associated with its product, as per the distribution code of practice [7]. In the absence of any regulations, this principle must still be followed. Had Sinochem been a RC company, this incident would contribute to its record as a non-self healing company, which are grounds for a failed verification and removal from the association [24].

2.4.1.2 The intermediary: VOS B.V. (now Helm Chemicals B.V.) in the Netherlands

While China is considered a developing country, the Netherlands is a member of the European Union and companies operating in this jurisdiction are subject to an array of legislation and voluntary measures, such as the European Chemical Industry Council's (Cefic's) RC program. VOS B.V., a wholly owned subsidiary of chemical giant, Helm AG, violated numerous product stewardship standards and principles. In terms of environmental management system checklists, the company followed protocol and sent a sample from the Chinese DEG-tainted containers to be tested. However, before receiving the results they shipped the containers to Haiti [17]. The company violated the first principle, which states that companies will "distribute chemical products and services in a manner which protects people and the environment from hazards." [7, p.33].

The company violated other codes of conduct when it failed to notify Haitian authorities of the toxic shipments. Furthermore, the company mislabeled the container as "98% pure USP", which they could not have known at the time of labeling, as the test results were received after the shipment left the dock for Haiti. RC maintains that a company should meet and exceed the law; however VOS B.V. did not follow this requirement and was prosecuted by the Dutch government and found guilty, with an imposed fine of $250,000 [17].

One downfall of previous European legislation before REACH was enacted was that it required manufacturers and importers of chemicals to provide information on hazardous chemicals, but did not require downstream users to do this unless the substance had to be classified and a safety data sheet had to be supplied with it further down the supply chain. According to the European Commission [19], "information on uses of substances was difficult to obtain and information about the exposure from downstream uses was generally scarce."

An RC company would be required to communicate information about their product and "obtain, understand and provide up-to-date material safety data sheets [and] provide to the customer information which the company believes to be vital to the health and safety of the end user" [7, p.35]. In the absence of cohesive legislation, the RC code is very clear that not only should the company abide by the law, but it should also go above the law to ensure the safety of the end user. Had VOS B.V. been a RC company it would have practiced due diligence and understood the principles of product stewardship. Since it did neither, VOS B.V. would fail its RC verification and be subject to reprisals [24].

The legal reprisals taken against VOS B.V. by the Dutch government would have been complimented by RC's structure, which imposes penalties for non-compliance while encouraging companies to improve themselves. The two mechanisms that achieve this are the verification process, which is conducted by an independent third party team and ensures that the company is in compliance with all RC directives; and the Chief Executive Officer (CEO) forum, which requires CEO's to sign off to the fact that their company is upholding all RC codes of practice. This forum exerts peer pressure on the CEO and effects his or her reputation, which could damage their corporate image, ability to trade with other companies and future employment potential [25].

In addition to these two mechanisms, community advisory panels and national advisory panels act as independent watchdogs of individual companies on a local level and of the industry association on the national level. These mechanisms work to ensure that the issues encountered by VOS B.V. would be avoided under RC's management [26].

2.4.1.3 The multinational firm: Helm AG, Germany

Helm AG has an intimate connection with the Haitian case study. Not only was it the parent of the wholly owned VOS B.V., it also owned one third of Sinochem at the time of the incident. Furthermore, the company has a history of mislabeling its products and was implicated in faking quality certificates for drugs delivered to Botswana [3]. A common theme in this case has been faked quality certificates and mislabelling, especially in products destined for third world countries. Numerous counterfeiting issues have also occurred in North America, but more rigorous testing requirements have averted disaster.

At the time of the incident, Helm AG was not a registered RC company, but it was certified under German and International Standards Organization (ISO) standards in 1992. The company considers itself a chemical marketing company, but has a reputation of falsely stamping products "Made in Germany" [3]. This inaccurate labelling, which would not be tolerable by any environmental management system, including ISO 14001, led the Haitian company to believe that the glycerin shipment came from European production facilities. The Helm AG example provides evidence that companies registered under environmental management systems can still evade checklist-based quality assurance measures. In later sections, it will be shown how RC combats this issue and complements checklists with ethic-based embedded philosophies.

According to the letter of the law, Helm AG was not responsible for the infractions made by VOS B.V. and therefore, was not punished by the United Nations Rapporteur [17]. However, a RC company who follows all principles and codes, practices self-regulation. According to the Codes of Practice, "self-regulation requires companies to meet all laws and regulations in spirit or exceed them. To achieve this demands ethical thinking, decision-making and performance" [7, p.6]. In 2000, around the time of the Rapporteur's investigation, Helm AG became a RC company and has been re-verified three times since [27].

Helm AG is representative of a retailer link in the supply chain which builds consumer trust and confidence with the products they sell. Retailers and producers need to trust the chemical sources that make up the products being retailed. At any point in the supply chain, a negative event can harm workers or users and damage a firm's reputation. A lack of transparent communication at any of links in the chain could cause considerable damage. In order for the system to remain safe, transparent and accountable, each partner in the chain must implement similar product stewardship strategies [5, 18].

2.4.1.4 The destination: Pharval in Haiti

The FDA investigation of the Pharval facility that produced the DEG-tainted cough syrup found that the facility did not have the proper health, safety, testing or clean-room areas in place. While this is disconcerting, had Pharval been a RC company, the situation would not have been avoided. Under RC, Pharval's health and safety measures would have been greatly improved, but their testing facility would still not have been able to test for diethylene glycol contamination, as it was too expensive and unavailable at the time [21]. In some respects, the company had no choice but to trust the label.

The laws governing the chemical industry in Haiti are sub-standard and their poor economy does not provide a good climate for research and development [21]. However, according to the RC Code, Pharval did not practice due diligence, because it did not take all the actions necessary to ensure that its product was safe for the end user. RC's research and development code of practice would have proven helpful, as it would have made Pharval aware of the hazards associated with their product and substandard facility [7].

Ultimately, in a developing country such as Haiti, it will be difficult for a single company to gain enough resources to implement the RC codes. An industry association of chemical companies may provide a means for collective action to improve laws and raise standards. One of the main principles of RC is to use its environmental and corporate responsibility codes to influence, guide and promote government regulation. In Canada, the CCPA helped influence the creation of Canada's National Pollutant Release Inventory which, along with other initiatives, is thought to have greatly increased the level of trust between government and industry in the country [25].

2.4.2 Theory 2: The loss of business, reputation and profits through erosion of trust was not a major consideration in the decision-making process of the companies involved

The public's perception of the chemical industry is very sensitive to chemical-related events and disasters regardless of where they occur and which parties are involved. In the past, the actions of non-RC companies in different parts of the world have affected companies financially, and in terms of their reputation. A high priority of the major chemical companies in the world is to protect the industry from negative events that threaten its stability [5, 25]. Therefore it is in their best interest to promote a responsible and safe industry worldwide, to avoid negative economic effects for companies who endure poor public perception, to alleviate customer fears of foreign products, and to mitigate business case effects associated with trust erosion and lack of regulation in regions where public scrutiny is not a motivating factor.

In the Haitian case, upon examining the actions of the companies involved, it is clear that they did not appreciate or fully understand the implications of their policies and behaviors. Sinochem, Helm AG, VOS B.V. and Pharval all suffered from erosion of trust. VOS B.V. and Pharval experienced this most directly in dealing with lawsuits and investigations; however, indirectly all companies involved in complex chemical supply chains suffered. DEG-tainted products resulting from poor supply-chain management techniques, erode the trust of users along the supply chain. "Recent business scandals have shaken public confidence in private corporations, increasing in turn the salience of principles of accountability, transparency, and integrity in all facets of business relationships" [28, p.1]. REACH legislation, American backlash against Chinese imports and attempts to strengthen RC globally have all arisen, in part, from an erosion of trust.

2.4.2.1 The business case for transparency and accountability in Responsible Care

Understanding the business case for sustainability is an important factor in motivating companies to engage in supply-chain management and corporate social responsibility. However, this requires that companies understand that a wide array of factors other than shareholder value can affect a company's financial performance.

The concept of corporate social responsibility is generally understood as a way to create business value from sustainable, ethical and philanthropic activities which allow a company "to meet and even exceed the legal, ethical and public societal expectations ... acting in a manner that respects the legitimate goals and demands of all stakeholders" [28, p.2].

Much of a company's willingness to accept sustainable principles depends on their developmental stage. Elkington [29] and Willard [30] both describe a variety of company types and stages of corporate strategic development. Corporate development can be broken down by types such as laggards, compliers, beyond compliance firms and model corporate citizens. Laggards and compliers generally do not understand or accept the business case, while the latter company types are working towards or are implementing the principles of green chemistry and sustainable development. Corporations in the first two stages of development may be enticed to conduct transparent operations if they were able to realize that a business case exists for a sustainable environmental management strategy [28, 30, 31]. The companies in the Haiti case were most likely in the laggard or complying stage of their development, which would explain why they ignored issues associated with the business case. If these corporations subscribed to RC they would have been encouraged to progress to the beyond compliance stage and avoid the events that occurred, since RC requires that complying companies adopt these principles [26].

When examining the business case, financial performance is not the only measure used to determine success. Shareholder value, revenue, operational efficiency, access to capital, customer attraction, the building of intellectual capital, status of a company's risk profile, innovation and social license to operate are all indicators of financial success and should be considered when examining the business case for sustainable development [32]. In order to accept that these additional measures are important considerations, corporate leaders must move beyond financial bottom line thinking and embrace a sustainable philosophy. Had the companies in the Haiti case understood this concept they would know that their attempts to falsify documents, hide information, counterfeit substances and deny responsibility could severely harm their social license to operate, increase their risk profile and decrease their ability to attract potential customers.

A RC company is required to prove its integrity and competence through the verification process, which ensures proof of implementation. The community advisory process seeks to engage all stakeholders in a meaningful and interactive way in order to gain the trust of the community and legitimize the community's trust in the company. Stakeholder engagement is an important element of fostering transparency, legitimacy, and stakeholder understanding of the various dangers associated with chemical manufacture. This philosophy recognizes that the public has the right to know how decisions are being made and a right to a traceable decision-making process ensuring accountability [33]. At the time of the Haiti incident, only Helm AG was registered under ISO 9001, which does not require these explicit elements of maintaining trust. The absence of these measures was a barrier to effective supply-chain management in the Haiti case [34].

2.4.2.2 Evidence of transparent and accountable business case success

Research conducted by the CCPA has shown that the business case for RC initiatives is very strong. The benefits are being marketed to firms as a way to reduce costs and obtain savings in efficiencies, reduced injuries and environmental clean-up, thereby increasing the firm's ability to influence the business environment, improve employee and community relations and lower insurance costs [26]. According to King and Lenox [35], RC's proactive approach can lower the costs of not complying and reduce waste production. This is cheaper than overproducing waste, makes environmental problems more manageable and can be powerful motivators in encouraging laggard companies to adopt RC principles and codes.

A firm that is open and transparent in its communication with stakeholders will allow it to anticipate community needs and improve manufacturing processes. According to Willard [30], a company can progress beyond financial motivators, where sustainability is a reward in itself. Although the concept of cradle-to-cradle and zero waste is not explicit in RC, evidence shows that RC companies are achieving some of these goals [36]. For instance, between 1999 and 2004, greenhouse gas emissions were reduced by 24% amongst CCPA companies even though production has increased [37]. The proactive supply-chain management philosophy of RC adds value for the end user and can create opportunities for synergy such as product take-back and recycling to reduce the costs of raw materials extraction.

Over the years, RC companies have been rewarded with lower insurance rates, lower financing rates, fewer legal issues, and better treatment by government agencies. A study by Earncliffe Strategy Group showed that in addition to these benefits, companies with a better reputation have lower overheads than companies with poor reputations. This is attributed to legal costs, corrective actions, erosion of employee morale, unwanted media attention and product de-selection [26].

2.4.2.3 Transparency, accountability as drivers for compliance and voluntary management

According to Lenox and Nash [38], the public tends to associate a chemical firm's performance with their industry rather than on an individual basis, making RC more lucrative for these visible chemical companies. Over the years, Helm AG became extremely visible and may have felt pressure from the chemical industry and customers to become more transparent and accountable. Just after the United Nations Rapporteur's investigation, Helm AG and VOS B.V. (now Helm Chemicals BV), became RC companies. Since then, Helm AG has garnered record profits [39] and teamed up with other RC companies, such as Nova Chemicals [40], to increase its operations worldwide. It would follow that by becoming a RC company and increasing its transparency and accountability, Helm AG and VOS B.V. decreased their environmental impact, increased their integrity and openness, and enjoyed financial success.

These notions of integrity and accountability tie into the concept of brand value which is an essential component to building a better business case. If RC companies are doing more to ease public concerns and transparently report on their operations, the resulting improved and more accountable relationship with stakeholders will have financial benefits by increasing brand value [26]. Regardless of whether a company follows guidelines and acts responsibly, they are often times judged on how they demonstrate their progress, how they interact with their stakeholders and what they are affiliated with [41, 42].

RC aims to demonstrate to stakeholders that member firms are responsible corporate citizens who create voluntary regulations that are dedicated to improvement. If the companies in the Haiti case subscribed to RC, they would have been motivated to follow corporate governance standards which promote due diligence. None of the non-RC companies in the case study were able to fully achieve these goals.

2.4.2.4 Risk assessment and management as drivers for compliance

The companies involved in the Haiti case did not incorporate the four cornerstones of trust in their daily transactions; these are integrity, openness, competence and empathy. These components are essential to risk management and profitability [33]. The fact that the Haiti supply-chain companies did not take these considerations into account helps to explain why the companies were inclined to counterfeit, forge and deceive. The evidence indicates that the companies failed to undertake a risk assessment which involves four major steps: hazard identification, dose response assessment, exposure assessment and risk characterization [43]. Once completing a cost-benefit analysis, it would become obvious that the costs of mismanagement would far outweigh the costs of compliance. A RC company internalises this concept, and uses it as a driving force to achieve positive financial, social and environmental outcomes [26]. Much of RC's Community Right to Know policy, which binds companies to be accountable to the public and balance risks and benefits, formally recognizes these elements of risk and trust [7]. At the core of RC are the trust building components of integrity, openness, competence and empathy, which were the aspects of the Haiti supply chain that were lacking or non-existent. Had the companies in the Haiti case recognized these elements as part of a risk assessment process, they could begin to mitigate and manage risks.

2.4.2.5 Challenges for Responsible Care and sustainable supply-chain management

Overall, the business case for RC presents challenges and opportunities for the chemical industry. The evidence indicates that firms who believe in the business case will conduct their business in a more transparent manner because they understand the financial, social and environmental benefits of sustainable development.

While it is evident that a strong business case for RC is accompanied by a strong motivation for firms to be transparent and accountable, in some countries, trust issues are not as important as they are in others, especially in those with controlled economies [34]. In addition, populations without access to an independent mass media may not be aware of supply-chain mismanagement and resulting health effects. According to Jerry Prout [44], global consumers and investors will come to expect environmental, ethical and social performance from multinational companies. RC can be adopted by companies which operate in countries with few regulations, in order to guarantee transparency and accountability.

In terms of RC's principles, Andy Smith from Earth Ethics suggests that the RC ethic must focus on sustainable development in its entirety. To do this, chemical firms must question the sustainability of the products they manufacture over their entire life cycle. The goals of zero-waste and zero-impact must be explicitly incorporated into RC, to ensure maximum transparency [45]. While efficient technologies can minimize environmental impacts they can also allow firms to manufacture higher volumes of hazardous chemicals at a reduced cost, allowing the chemical to flourish and drive consumption despite its negative environmental effect [31]. A sustainable strategy would examine the social aspects of hazardous chemicals, including health and safety issues, and would pursue a policy of reducing or replacing these chemicals with more benign and cost effective alternatives [45]. This type of strategy is one that maintains a congruency between policies and ethics. Chemical firms need to resolve their multiple interests and strive to achieve implementations that correspond to ethical goals prescribed by RC [25].

Reporting on economic, social and environmental indicators is considered an effective way to ensure transparency across a variety of issues [29]. The CCPA monitors some of these indicators but not all of them and other jurisdictions are far behind the Canadians. According to Andy Smith "the industry has not achieved effective and transparent self-assessment" [45, p.44]. Industry associations under the guidance of the ICCA and their Global Product Strategy are being encouraged to adopt high RC standards such as those set by the CCPA [10].

The situation is further complicated by lies of omission. Companies who report on environmental, health and safety measures sometimes choose not to report on indicators they have performed poorly on. Without standardized reporting indicators and procedures, reports tend to be green wash rather than a factual representation of a company's performance [46, 47].

In a 2003 review of the CCPA's activities, the National Advisory Panel (NAP) cited problems which could call into question the credibility of the third party RC evaluations. The CCPA has allowed some laggard firms to continuously postpone their evaluation dates. According to the panel, for the system to remain credible and accountable, action and sanctioning must take place to ensure that member firms meet their expectations [48].

The requirement of an independent, third party evaluation is an important one; it helps to increase transparency and provides a mechanism to monitor progress. However, without sanctions, non-compliant member firms can still free ride on compliant ones. This also applies to product stewardship and supply-chain management. Partner firms are more likely to comply, if they are under the threat of sanctions for non-compliance. "Studies indicate that [voluntary environmental programs] without sanctions, independent oversight, and standards ... are not effective in promoting improved corporate environmental performance." [49]. One possible sanction is to publish the results of verifications, which the CCPA already does, and set targets for emissions reduction per company [25]. If a company does not meet emissions targets, their poor performance is published on a website.

In the near future RC and the Global Product Strategy will have to ensure that the principles of green chemistry are used to implement risk reduction measures using precautionary, rather than reactionary policies. Simply identifying risks without managing them is not enough to avoid future supply-chain management issues that go beyond what happened in Haiti. The actions by the companies in the Haiti supply chain were deliberate and blatant; however, in the future, dangerous chemicals used by responsible companies could have unknown risks with possible negative environmental, social and health effects [43].

2.4.3 Theory 3: The companies involved in the case study did not embed the philosophies of Responsible Care ethic or create a corporate culture of stakeholder protection

In the analysis of supply-chain management issues in the Haiti case, RC was mainly referred to as an environmental management system. However, RC is an ethic and philosophy, as much as it is a management system. This duality of RC is what provides it with the ability to be self-critical and evolutionary [5]. The analysis of RC as ethic will help to explain why Helm AG and its subsidiary, both ISO registered companies, were capable of such blatant negligence and deceit.

2.4.3.1 Embeddedness: Responsible Care as ethic

Integrating RC into the corporate mandate is a major focus of chemical industry associations. This is highly evident in the fact that one of the requirements of the verification process is that the RC ethic be evidenced at every level of the corporation [24].

Although RC is more than a health and safety initiative, many critics treat it as such, especially those companies who have not been exposed to the benefits of sustainable development [50]. They tend to relegate RC initiatives to the company's environmental, health and safety department. According to Willard [30] and Hawken [31], compartmentalizing sustainability ethics will not result in superior environmental performance, nor will it lead to an increase in efficiencies or profits. Furthermore, this type of compartmentalization does not integrate sustainable thinking into the culture of the organization and does not allow the company to obtain the financial and social benefits it could achieve with a truly transparent and accountable mandate [30].

Firms, such as those in the Haiti supply chain who did not implement RC initiatives, performed more poorly and non-transparently. The companies had environmental management systems but they did not integrate the philosophy of sustainability and care into their corporate culture.

2.4.3.2 Ethics and incentives for over-compliance: Going beyond environmental management

According to King and Lennox [35], there exists a possibility for conformity without sanctions through three mechanisms: coercive forces, normative forces and mimetic forces. Coercive forces involve damaging a firm's reputation in front of its peers and the general public. This type of strategy puts pressure on lagging companies through public inquiry or peer reprimand and shaming. RC provides this through regional leadership groups and community advisory groups. These groups work to ensure that companies embed the ethic and comply with the management system. It also helps to alleviate fears that some companies may agree to RC but then ignore their responsibilities [25].

Normative forces include the adoption of external values into the existing corporate culture. In terms of RC, corporations who take on the initiative begin to adopt its values into their mission, vision and day-to-day operations. "Responsible Care is ... a process woven into the fabric of a company's culture ... It has to become part of the way it operates so it becomes ingrained in the company's day-to-day operations" [7 p.3].

Mimetic forces enable a company to learn from its peer companies through social networks established through its industry association. This strategy requires transparency, accountability and communication on the part of all participating firms. RC encourages firms to share their best practices for environmental, health and safety performance [7, 51].

Coercive forces were not present in the environmental management systems that the companies in the Haiti case subscribed to, while others had no Environmental

Management System (EMS) in place at all. ISO 9000 and 14000 for quality and environmental management respectively, provide an excellent system to improve processes using prescriptive measures and checklists. RC's added ethical dimension sets it apart from these standards, particularly with respect to the independent verification of ethical aspects of the organization's operations. These additional factors go above and beyond typical EMSs and would have been instrumental in deterring the companies in the Haiti supply chain from engaging in unacceptable behavior [34].

2.4.3.3 Responsible Care as a compliment to legislation

Regulations are important and necessary tools in developing sustainable supply chains; however, they are not designed to encourage firms to perform beyond that which is mandated by the law [30, 52]. According to Willard [30], regulations need to be augmented by Voluntary Environmental Programs (VEPs), government incentives and prevention strategies in order to achieve results that increase company performance beyond that which is mandated by the law. VEPs should not be a replacement for regulations, nor should regulations preclude the use of VEPs. The threat of regulation combined with facilitated partnerships can encourage manufacturers to develop stringent voluntary programs which keep their governance policies ahead of regulations and allow for more flexible mechanisms to achieve their goals [52].

The Environmental Protection Agency's Toxic Release Inventory and the European Union's REACH are two examples of initiatives which overlap with RC directives. The REACH initiative focuses on product stewardship and makes the chemical producers and suppliers responsible for hazardous chemicals during throughout their life cycles. According to the chairman of the ICCA's RC leadership group [10], this type of registration and supply-chain management has been an un-fulfilled goal of RC for many years. The Cefic has also revamped its product stewardship directives in order to meet and surpass the new REACH legislation.

Although REACH requires information gathering and testing of dangerous chemicals, it also requires companies to be responsible for risk reduction measures. This requires company self-responsibility in order to implement risk management procedures. Control and sanctioning mechanisms are integral to self-responsibility but can be hard to enforce. If this is not implemented properly "REACH will collect data about risk information without significantly forcing or encouraging risk reduction measures" [43, p.40]. RC can fill this gap and encourage member companies to take responsibility for the chemicals that they produce while ensuring that the risk information collected by firms is comprehensive and performed with due diligence.

The laggard companies of today may not respond well to legislation such as REACH on its own. This was the case with Helm AG before it signed on to the RC program. When regulatory and voluntary systems are functional and properly enforced their synergy can achieve the best results in "beyond compliance" corporate behaviour [30]. This addresses the fact that legislation cannot always keep up with technical innovation.

This gap-filling hypothesis posits that not only would the Haiti companies fail verification, the CEO and board would be held responsible and receive reprimand and guidance from the committee of industry peers who all stand to have their reputations tarnished by the failures of the weakest links. None of this could be achieved by legislation alone; however, in reality this process could break down or not be followed correctly by those implementing the sanctions. In addition, the verification process only takes place once every three years so it is not effective as an immediate form of sanctioning. It may also be difficult to determine whether companies change their behavior after an incident because of RC sanctioning and ethics or because of some other factor such as public pressure and outrage [41].

While this hypothesis could be proven wrong in practice, legislation alone cannot help foster inter-firm communication, nor can it develop mutual support networks for chemical firms. These social structures help relate information about best-practices to all the firms in a given association. The initiative also encourages peer review of reporting data and mutual assistance when required, to increase compliance and competitiveness [41].

The companies involved in the Haiti case would have benefited from the RC's encouragement to move beyond compliance, a philosophy which has the potential to help laggard companies avoid the issues encountered in the Haiti supply chain.

2.4.3.4 Corporate culture and international supply chains

Embedding a culture of protection in the supply chain is a focus of RC which gives it a central role in providing a way to maintain consistency in regulation between countries. It provides an international solution by addressing the root of supply-chain management issues which threaten the chemical industry [53]. RC companies integrate a Plan-Do-Check-Act management system which incorporates international product stewardship guidelines as well as green chemistry principles implicitly. Laggard companies who adopt RC principles, such as those in the Haiti case, would come to internalize these concepts into their corporate culture to achieve the successes that many RC companies share.

Research by Willard [30] and Senge et al. [54] demonstrate that only a culture change can bring about legitimate results that proliferate throughout the whole company. This concerns all the levels of an organization from CEO to employees. Employee buy-in to the environmental management system and social ethic of the company is viewed as an essential element in creating a culture of integrity [55].

Reviewing the case of the Haiti supply chain reveals that this corporate culture did not exist. Had there been a culture of ethical concern, drug counterfeiting would not be tolerated by the company or the employees that allowed it to occur. This is also evidenced by examining the actions of RC companies which have embedded a

Table 2.2: Summary of infractions and results in the Haiti case study

Company	Infraction	Result
Sinochem (China)	– Produced counterfeit glycerin tainted with DEG – Improperly labelled the shipment as 98 PCT USP – Faked the Certificate of Authenticity	– No actions taken against the company – Trackback traceability to implicate the company was not possible – Chinese government has strengthened pharmaceutical laws
Helm AG (Germany)	– Packaged the product with German Authenticity and Made in Germany labels – Parent company of VOS B.V. – Owned part of Sinochem – History of Mislabelling products	– Not implicated in the legal investigation because it was not directly responsible – Claimed no responsibility for the case in Haiti – Became a RC Company, engaged in corporate social responsibility and has posted record profits
VOS B.V. (The Netherlands) – wholly owned subsidiary of Helm AG	– Failed to properly test Chinese shipment – Improperly labelled the shipment as 98 PCT USP by copying the shipped label – Used the fake Certificate of Authenticity – Knowingly transported the tainted counterfeit glycerin to Haiti	– UN Rappateur found VOS B.V. guilty of shipping the tainted product – Dutch Government found VOS B.V. guilty and fined them $250,000 USD – Settled a civil suit with Pharval and the affected families for $10,000 USD per family – Since the infraction, VOS B.V. is now a RC Company
Pharval (Haiti)	– Failed to properly test product – Unknowingly sold tainted product in the cough syrup medication – 85 children died of the poisoning as a result of their lack of due diligence	– Lost the trust and business of their customers – Settled in a civil suit with the families of the deceased children for $10,000 USD per family

culture of protection into their day-to-day operations. RC requires that companies operating in Canada ensure that RC codes are being followed in other countries and other companies along the supply chain. If RC or legislation is absent from the country in which the company's operations are occurring then the CCPA's version of RC will apply [24].

2.5 Conclusion and discussion

Global supply chains, stakeholder trust, due diligence and corporate culture are the key topics examined in the context of the three main theories which lead to the conclusion that RC, as implemented by the CCPA, would have mitigated or averted the crisis. These are:

- The companies involved did not internalize the concepts of product stewardship and the cradle to cradle philosophy that Responsible Care advocates.
- The loss of business and reputation through erosion of trust was not a major consideration in the decision-making processes of the companies involved.
- The major players involved in the case study did not embed the philosophies of the Responsible Care ethic or create a corporate culture of protection of all stakeholders.

RC needs to better promote preventative, rather than reactive approaches to risk assessment and risk management. RC's product stewardship principles can do more to encourage the substitution of the most dangerous chemicals. This approach would better demonstrate the industry's commitment to its principles and avoid claims that the industry has incongruent political and environmental goals. A related opportunity for the improvement of RC is the full integration of the principles of sustainable development into its mandate. A move to a more transparent and accountable process would most likely include an emphasis on sustainability and the principles of industrial ecology.

Overall, RC's commitment to continuous advancement will help capitalize on the opportunities for improvement in the areas identified and will continue to ensure safe and equitable global supply chains in the future.

References

[1] Rentz, E. D., Lewis, L., Mujica, O. J., Barr, D. B., Schier, J. G., Weerasekera, G., Kuklenyik, P., McGeehin, M., Osterloh, J., Wamsley, J. Lum, W., Alleyne, C. Sosa, N. Motta, J. Rubin, C. Outbreak of acute renal failure in Panama in 2006: a case-control study. Bulletin of the World Health Organization 2008;86:749–56.
[2] Centre for Disease Control. Fatal poisoning among young children from diethylene glycol-contaminated acetaminophen – Nigeria, 2008–2009. Morbidity and Mortality Weekly Report 2009;58(48):1345–1347.
[3] Schaaber, J., Jenkes, C., Wagner, C. Counterfeit medicines: What are the problems? Pharma-brief special: No. 1, BUKO Pharma-Kampagne, Bielefeld, Germany, 2007.
[4] Guidance for industry: Testing of glycerin for diethylene glycol. Food and Drug Administration, Centre for Drug Evaluation and Research, Maryland, USA, 2007.
[5] Bélanger, J. M. Responsible care in Canada: The evolution of an ethic and a commitment. Chemistry International 2005;27(2):1.
[6] Leblanc, H., Milek, F. Good pharmaceutical trade and distribution practices. WHO Drug Information 2001;15(1):2–5.
[7] The ethic and codes of practice of responsible care. Canadian Chemical Producer's Association, Ottawa, 2000.
[8] Beamon, B. M. Supply chain design and analysis: Models and methods. International Journal of Production Economics 1998;55:281–294.
[9] The Lean and Green Supply Chain. EPA 742-R-00-001. Office of Pollution Prevention and Toxics, Environmental Protection Agency, Washington DC, USA, 2000.

[10] Product stewardship guidelines. International Council of Chemistry Associations, Arlington, VA, 2007.

[11] McDonough, W., Braungart, M., Anastas, P. T., Zimmerman, J. B. Applying the principles of green engineering to cradle-to-cradle design. Environmental Science Technology 2003;37(23):A434.

[12] Kainuma, Y., Tawara, N. A multiple attribute utility theory approach to Lean and Green Supply Chain Management. International Journal of Production Economics 2006;101:99–108.

[13] O'Brien, K. L., Selaniko, J. D., Hecdivert, C., et al. Epidemic of pediatric deaths from acute renal failure caused by eiethylene glycol poisoning. Journal of the American Medical Association 1998;279(15):1175–1180.

[14] Buddingh, H. The killer cough syrup. World Press Review, May, 1997.

[15] Kenyon, A. S., Xiaoye, S., Yan, W., Har, N. G. Simple, at-site detection of diethylene glycol/ethylene glycol contamination of glycerin and glycerin-based raw materials by thin-layer chromatrography. Journal of AOAC International 1998;81(1):44–50.

[16] Bogdanich, W. F.D.A. tracked tainted drugs, but lost trail in china. New York Times, June, 2007.

[17] Ksentini, F. Adverse effects of the illicit movement and dumping of toxic and dangerous products and wastes on the enjoyment of human rights. Fifty-fifth session, item 10 of the provisional agenda No. E/CN.4/1999/46. United Nations Economic and Social Council: Commission on Human Rights, Geneva, 2000.

[18] Ten Ham, M. Health risks of counterfeit pharmaceuticals. Drug Safety 2003;26(14):991–997.

[19] REACH in brief. Environment Directorate General, European Commission. Brussels, Belgium, 2007.

[20] Quality assurance of pharmaceuticals. Good Manufacturing Practices and Inspection. World Health Organization, Geneva, 1999, 2.

[21] Junod, S. W. Diethylene glycol deaths in Haiti. Public Health Reports 2000;115(1):78.

[22] Beach, M. China opens drug market by revising pharmaceutical law. The Lancet 2001;357(9260):942.

[23] Drug safety and availability - FDA initiative to combat counterfeit drugs. Food and Drug Administration, 2009. (Accessed November 28, 2010, at http://www.fda.gov/Drugs/DrugSafety/ucm180899.htm.)

[24] Responsible care re-verification 2005–2008. Canadian Chemical Producer's Association, Ottawa, 2008.

[25] Moffet, J., Bregha, F., Middelkoop, M. J. Responsible Care: A case study of a voluntary environmental initiative. In: Webb, K., ed. Voluntary codes: Private governance, the public interest and innovation. Carleton Research Unit for Innovation, Science and Environment, Ottawa, 2004, 177–208.

[26] Does Responsible Care Pay? Canadian Chemical Producer's Association, Ottawa, 2005.

[27] Responsible Care verification certificate. The German Association of Chemical Trade and Distribution, 2011. (Accessed October 14, 2013, at http://www.hit.helmag.com/en/qualitymanagement/Responsible_Care_english.PDF.)

[28] Jamali, D. The case for strategic corporate social responsibility in developing countries. Business and Society Review (00453609) 2007;112(1):1–27.

[29] Elkington, J. Enter the triple bottom line. In: Henriques, A., Richardson J., eds. The triple bottom line: Does it all add up? Earthscan, London, 2004, pp. 1–16.

[30] Willard, B. The next sustainability wave: Building boardroom buy-in. New Society Publishers, Gabriola Island, BC, 2005.

[31] Hawken, P., Lovins, A., Lovins, L. H. Natural capitalism. Little, Brown & Company, New York, 1999.

[32] Dudok van Heel, O. Buried treasure: Uncovering the business case for corporate sustainability. SustainAbility, London, 2001.

[33] Baillard, V., Belanger, J., Steinberg, S., Dinsdale, G., Giroux, K. Building trust: A foundation of risk management. Canada School of Public Service, Ottawa, 2001. (Accessed August 15, 2008, at http://www.csps-efpc. gc.ca/research/publications/pdfs/Risk-Trust-REV.PDF.)

[34] B. Wastle. Personal communication. May 27, 2008 (details available on request from the authors).

[35] King, A. A., Lenox, M. J. Industry self-regulation without sanctions: The chemical industry's responsible care program. Academy of Management Journal 2000;43(4):698–716.

[36] Livesey, S. M. The discourse of the middle ground: Citizen shell commits to sustainable development. Management Communication Quarterly 2002;15(3):313–349.

[37] Kamalick, J. Responsible care set for improvements. ICIS Chemical Business Americas 2007;271(5):28–29.

[38] Lenox, M. J., Nash, J. Industry self-regulation and adverse selection: A comparison across four trade association programs. Business Strategy and the Environment 2003;12(6):343–356. (Accessed August 15, 2008, at http://www3.interscience.wiley.com/cgi-bin/fulltext/106560527/PDFSTART.)

[39] Press release – Helm Group 2012. Helm AG, 2013. (Accessed October 14, 2013, at http://www.helmag.com/en/news/news/press-releases-detail/news/press_release_helm_group_2013_review_and_outlook/.)

[40] Helm AG to market NOVA chemicals styrene monomer in europe. NOVA Chemicals, 2003. (Accessed June 22, 2008, at http://www.novachem.com/appl/prelease/news.cfm?id#equal#191.)

[41] Two decades of responsible care: Credible response or comfort blanket? Environmental Data Services (ENDS) Ltd., No. 360, London, 2005. (Accessed March 25, 2007, at http://www.responsiblecare.org/filebank/ENDSFeaturesResp_Care.pdf.)

[42] Reisch M. Responsible Care: The chemical industry has changed its ways, but more radical actions may be needed to change public opinion. CENEAR 2000;78(36):21–26. (Accessed August 15, 2008, at http://pubs.acs.org/cen/coverstory/7836/7836bus1.html.)

[43] Koch, L., Ashford, N. A. Rethinking the role of information in chemicals policy: Implications for TSCA and REACH. Journal of Cleaner Production 2006;14(1):31–46.

[44] Prout, J. Corporate responsibility in the global economy: A business case. Society and Business Review 2006;1(2):184.

[45] Schmitt, B. Industry's critics make respect an elusive goal. (Responsible Care). Chemical Week 2002;164(27):42(3)–45.

[46] Johnson, G. Don't be fooled 2005: America's ten worst greenswashers. The Green Life, Boston, 2005.

[47] Tschopp, D. J. Corporate social responsibility: A comparison between the united states and the European Union. Corporate Social Responsibility and Environment Management 2005;12:55–59.

[48] Message for 2002/2003 responsible care report. Canadian Chemical Producer's Association (National Advisory Panel). Ottawa, 2003.

[49] Steelman, T. A., Rivera, J. Voluntary environmental programs in the united states: Whose interests are served? Organization & Environment 2006;19(4):505–526.

[50] Givel, M. Motivation of chemical industry social responsibility through responsible care. Health Policy 2007;81(1):85–92.

[51] Prakash, A. Responsible Care: An assessment. Business and Society 2000;39(2):183–209.

[52] Hoffman, A. J., Riley, H. C., Troast, J. G., Jr., Bazerman, M. H. Cognitive and institutional barriers to new forms of cooperation on environmental protection: Insights from project XL and habitat conservation plans. American Behavioral Scientist 2002;45(5):820–845.

[53] The responsible care initiative: Lessons learned. Canadian Chemical Producer's Association, Ottawa, 2002.

[54] Senge, P., Dow, M., Neath, G. Learning together: New partnerships for new times. Corporate Governance 2006;6(4):420.

[55] Corporate responsibility: Private initiatives and public goals. Organisation for Economic Co-operation and Development, Paris, 2001.

Peter Topalovic, Maria Topalovic, and Gail Krantzberg

3 Responsible Care's effectiveness in promoting sustainable industrial performance: A case study in sustainable chemical production and distribution

Abstract: The main purpose of this study is to determine whether Sulco's environmental, social, and community engagement successes could have been achieved without the use of Responsible Care (RC). The central question of the study asks, *"Can a company that does not internalize the values of Corporate Social Responsibility (CSR), effectively implement RC codes and practices?"* This analysis focuses on stakeholder interviews, production data, and RC verification reports to determine the relationship between CSR, resiliency thinking, industrial ecology, sustainability, and RC. The analysis demonstrates that the corporate adoption of self-regulatory principles, policies, and programs can be effective in producing a positive impact on all stakeholders, including the community, environment, employees, and residents, if the company internalizes the values of CSR and its complementary concepts. The Sulco case demonstrates that when production data, community relations, and comparison with companies that do not internalize RC are taken into account, it is evident that a part of Sulco's success can be directly attributable to their participation in the RC program. The success attributable to RC is only possible when the company internalizes sustainability principles and does not subscribe to RC only to greenwash.

3.1 Overview and background

3.1.1 Sulco Chemicals case study

Sulco Chemicals Limited (Sulco) is based in Elmira, Ontario with 20 employees. It is owned by Canada Colors and Chemicals Limited (CCC), now branded as the CCC-Group, which was established in the 1920s and is now one of the largest chemical distributors in North America. While this study focuses on the Elmira subsidiary, CCC ethics will also be investigated. Operating in a small footprint with a modest staff complement, Sulco manufactures 80,000 metric tonnes of sulfuric acid, oleum and SBS with by-product steam production. It also packages the sulfuric acid and SBS, as well as hydrochloric acid, aluminum sulfate, ferric chloride, hydrofluoric acid, caustic soda, nitric acid, and hydrofluosilicic acid.

From an environmental standpoint, the plant has steadily reduced stack emissions by 96% of 1990 levels by continuously improving plant processes. Additionally,

Sulco has found markets for its various by-products to maximize the use of feedstocks into the local market. Sulco has successfully engaged the surrounding community members through its Community Advisory Panel (CAP), worked with the community to improve their safety in the event of an emergency, integrated community feedback and suggestions in plant operations, and reduced emissions to improve air quality and reduce greenhouse gas emissions [1].

In the late 1980s there were four "chemical" plants in Elmira, each of which had not fared as well with health and safety issues, public engagement, and emissions mitigation. Two of the plants have closed, and the remaining one has continued to struggle with the issues, causing Sulco to stand out amongst its peers and be recognized by the community as a corporate leader in RC [2].

The town of Elmira is located in Woolwich Township, in the region of Waterloo, with a population of 9,900 [3]. Much of the industry is focused on Union Street and the area adjacent to Canagagigue Creek. Since the early 1990s, the Elmira Aquifer System has been deemed too contaminated to supply drinking water to the town's inhabitants and in a settlement with Uniroyal/Chemtura, the major polluter in the region, and the Ministry of the Environment (MOE) [4], a control order was issued to remediate the aquifer to clean drinking water standards by 2028. Elmira residents currently have their water supplied by pipeline from Waterloo Region.

3.1.2 Central question and study purpose

The main purpose of this study is to determine whether Sulco's environmental, social, and community engagement successes could have been achieved without the use of RC. This chapter is to develop a case study that highlights a successful implementation of RC and is part of a series of cases for the IUPAC projects on RC. It is possible that a further case could be developed to compare the approaches and issues of other Elmira companies who have faced challenges in the past.

Much of the previous research on RC evaluates its historical development [5], its effectiveness [6–12], or its legitimacy [13–15]. This study recognizes that RC has genuine impact and effectiveness, but examines the effect of context on the outcomes of RC. It asks the question, "Does implementing RC contribute to better environmental, economic, and social outcomes?" This includes Corporate Social Responsibility (CSR) which is a self-regulatory set of principles, policies, and programs that a corporation adopts to ensure that the company has a positive impact on all stakeholders, including the community, environment, employees, and residents. It will also examine to what extent RC has impacted the company in terms of community profile, economic considerations, and environmental performance, especially when compared to ISO 14001 (which the company is also registered under) and government regulations, which it meets or exceeds.

Throughout the analysis, the degree to which RC has been integrated into the day-to-day operations of the company at both Sulco and CCC will be assessed. This aspect of the study will focus on the RC verification results, because they provide key information on the level of RC integration. The focus of this study is on Sulco, and the data collected will focus mainly on Sulco data.

This timely research can help examine the positive implementation of RC and whether Sulco's success is a result of the application of RC codes and principles or due to regulatory compliance, ISO certification, or good management practices. The research being conducted in the case study series has several important implications for companies in the industry, for communities located near chemical plants, and for the country as a whole, where the ecosystem can suffer from large emissions or disaster-related impacts. If RC can deliver the claimed benefits that are being investigated, it can have a significant impact that other measures, systems, regulations, and programs cannot achieve [5, 6, 8–10, 16].

A RC program that is proven to have real community value can help assure residents that their health, safety, and wellness are being considered as part of regular plant operations. It should encourage residents and municipalities to apply pressure to non-RC companies to become verified and demonstrate positive results through the re-verification process. From a national policy perspective, encouraging or requiring large emitters to have effective industry-led management programs in place may help to improve air quality and reduce greenhouse gas emissions and other environmental impacts.

3.2 Methods

The data for this research is drawn from a variety of sources. Sulco's emissions and plant process information was obtained directly from company data and the National Pollutant Release Inventory (NPRI) data [1]. This provides evidence of continued improvement and emissions data that go above and beyond regulations, as evidenced in Fig. 3.2: Sulco's sulfur dioxide (SO_2) emissions (1990 to date). This data was compared to the amount of input materials used, as a measure of production scale, in order to ensure that not only was the process becoming more efficient, but as the company increased their production, they were decreasing their actual tonnage of emissions released.

Another important source of data, which was used to analyze the level that RC was integrated into corporate culture and processes, was the verification report documents for Sulco [21, 24–26]. These reports were analyzed for the adherence to RC codes and practices, environmental and social impact, level of RC ethic saturation into all aspects of the business and plant operations, accountability, process management, and commitment to sustainability. These reviews provide evidence and insight into the use of RC and its effectiveness, assisting in the evaluation of RC as it pertains to the central question of this research. It is important to highlight that the Canadian RC program is built on a foundation of third party verification, which is not the case in all 53 countries that have RC programs; although countries such as the United States have moved to this type of system [8]. King and Lenox, provide evidence supporting the fact that industry-led self-assessment schemes may not be effective without sanctions, which third-party verifications help to provide [14]. A company which is not verified cannot demonstrate to the public that it is self-healing, transparent, or meeting its commitment to the community [5].

Finally, interviews and surveys were conducted with representatives of the Sulco plant team, RC verifiers, members of the CAP, and Elmira residents to develop a multi-stakeholder assessment of RC at the Sulco plant and its effectiveness. The interview data provide key information on community relations and impact and the level of RC understanding in the plant and community. A standard question list was asked of each participant. These can be grouped into the following categories: (a) measure of RC's influence in Sulco's successes and community relations, (b) comparing and contrasting RC's impact to the impact of government regulations and other environmental management systems, (c) comparing Sulco's impact and stakeholder relations to those of neighboring plants, (d) determining the importance and success of verification and reporting. In total, ten interviews were conducted with representatives from each target group identified, and their responses were analyzed and compared. Due to the highly qualitative nature of this case study, the information collected through these interviews is an important element of the outcomes of this study.

The emissions data aids in the analysis of the company's environmental impact; the verification data analysis demonstrates the level of RC's effect on the company and its integration into the corporate culture; and the interview responses help gauge the social impact of the company and provide first-hand experience with the company's processes, business plans and philosophies.

3.3 Results

3.3.1 Responsible Care and its influence on Sulco plant operations

CCC was one of the early adopters of RC when it was developed in Canada in 1985 by the CCPA. In part, the initiative aimed to demonstrate to the public the willingness of the industry to protect the public interest; however, another important dimension was to ensure that all member businesses developed and sustained a management system that resulted in environmental, social, and financial success that exceeded regulations and projections. While CCC is primarily a chemical distributor, its subsidiary, Sulco, which is now known as CCC Sulphur Products, is a chemical producer. According to the former Chief Operating Officer, Bernard West [17], CCC had the option to adopt RC for distributors, which uses a subset of the RC codes that pertain to distribution, but instead they chose to subscribe to all the codes and practices of RC. It is clear that Sulco is an early adopter of good manufacturing practices and at the very least would like to communicate and promote that they support these codes. However, this study is concerned with what impact RC has had on the success of Sulco, over and above Sulco's commitment to sustainability, community, and good manufacturing practices that is inherent in the company's own ethic. Since the company has been involved with RC for many years, it will be necessary to understand historically how the company developed and what they integrated as a result of adopting RC. Fig. 3.1 shows the various elements and programs that contribute to process and product improvements as well as community engagement.

RC was designed to be an ethic that is internalized by all elements of the organization including processes, people, and profits. It provides policies, procedures, and strategies to improve the financial, administrative, environmental, and social elements of the company's operations. However, RC alone cannot be the sole catalyst behind all plant improvements. Regulations, other industry standards, the community, and the business case all impact a plant's operations [8]. ISO 14001 is an example of another voluntary industry program that audits technical procedures and ensures they are helping to meet identified environmental targets. This process-based environmental management system focuses in achieving process targets, whereas RC, an ethic based system, attempts to promote improvements in community engagement, environmental impact assessment, and emergency preparedness in addition to process improvements [16]. Both of these programs encourage continuous improvement, and according to the Vice President Ron Koniuch [18], Sulco required both to achieve their goals. While RC is specific to the chemical industry, ISO 14001 applies generally to the manufacturing process.

Regulations also play a key role in ensuring various industries are compliant with the minimum emissions standards and environmental targets, however, they are prescriptive and require enforcement. Some of those interviewed, including CAP

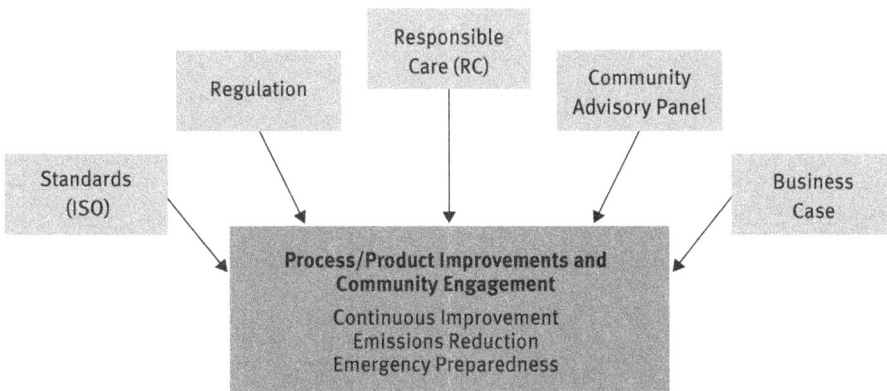

Fig. 3.1: Elements and influencers of process and product improvements

member Richard Clausi [2], expressed concern that the Ministry of the Environment has not always been able to properly enforce regulations. RC polices and verifications encourage companies to exceed regulations in order to internalize enviro-social standards. Sulco has always been committed to reducing emissions, and according to Fig. 3.2, in 1990, they produced 870 tonnes of sulfur dioxide, by 2011, they produced about 30 tonnes, which is 3.4% of the legal limit. The maximum allowable limit, according to the MOE is 830 metric tonnes (MT).

Woolwich Fire Chief Kieran Kelly [19] recalls the Elmira environment in the late 1980s (as it relates to the Varnicolour, Chemtura, Nutrite, and Sulco plant sites). At that time, all four companies had issues on both water and air emissions, however Sulco was the only company that met proactively with the Fire Department and Township officials very early on to discuss corrective actions for its plant operations. This attitude has continued with its origins in the same timeframe of the RC implementation process.

A review of the company's NPRI data demonstrates that Sulco's on-site air emissions have continued to decrease overtime. Sulco has reported on the following substances: hydrochloric acid, hydrogen fluoride, and sulfuric acid since 1994; formic acid since 1999; and sulfur dioxide since 2002. Both formic acid and hydrogen fluoride have been reported as zero emissions since 2000, resulting in a 100% decrease. Hydrochloric acid has decreased by 90% since 1994 levels, sulfuric acid by 35% since 1994 levels, and sulfur dioxide by 94% since 2002 levels [1]. The most evident change was the decrease in sulfur dioxide emissions from 2008 to 2009, which can be attributed to the installation of a scrubber that reduces sulfur dioxide emissions. The installation of the scrubber was due to a recommendation from the CAP, which is

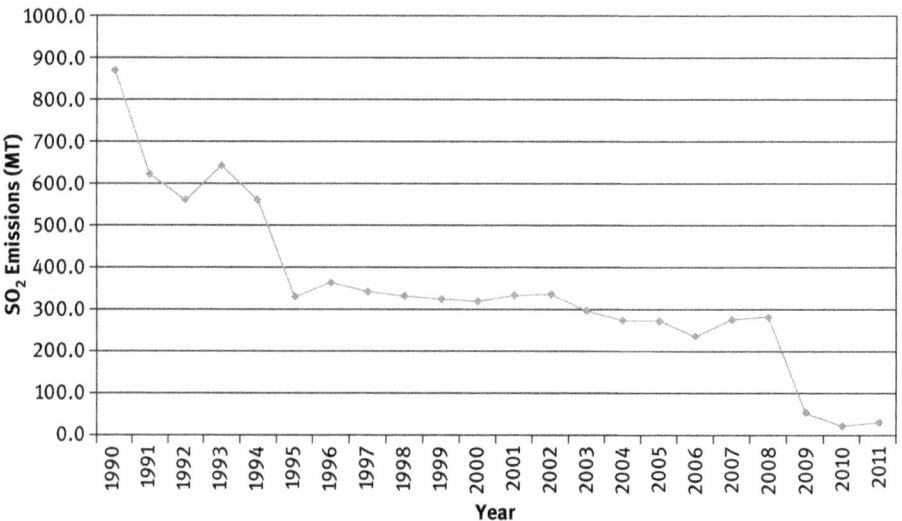

Fig. 3.2: Sulco's sulfur dioxide (SO_2) emission (1990 to date)

ultimately a result of Sulco's adoption of RC. However, it is important to understand how the interaction of RC, the CAP, and Sulco's business interest led to their positive emissions-reduction results.

In 2000, Sulco developed an environmental plan, in consultation with the community, which aimed to reduce emissions, save money, improve plant efficiency, and find new markets for by-products to increase profits. Another goal of this plan was to increase production without increasing emissions [18]. This was driven by the integration of RC codes and practices as part of the company's integration of RC into the company's processes. From the data examined it is evident that Sulco was able to greatly reduce emissions, while increasing production. According to Jean Bélanger, a Sulco Board member, emission reductions beyond regulations is reflective of the company's commitment to sustainability [20]:

> "A compliance mentality, in the long term, will always produce the minimum results which, at times, may meet the needs, but mostly, it will result in continuous apparent needs to tighten the regulations, probably requiring inefficient operational adjustments at higher costs. It is always more efficient and economical to determine what is the right thing to do and then to do it right, once and for all."

The comments made by Bélanger also reflect the company's goal to increase profitability through sustainable design. These environmental and social goals permeate from the top management team and ethic of the parent company, CCC, as well as from the employees on the shop floor. As an example, Sulco reduced greenhouse gas emissions by recuperating waste heat from one of their processes and selling the energy to another company for use in that company's process. The plant generally has low carbon dioxide (CO_2) emissions. The process uses stove oil for energy during start up and the office is heated electrically. Steam is generated in the process, which they sell to another company that reduces the need for that company to burn fuel to generate steam [21].

The sustainable behaviors exhibited by the company can be attributed to the business case associated with the profits that were realized from plant improvements. However, from the interview data collected [2], both company representatives and CAP members attribute the low emissions to RC and its emphasis on integrating community concerns into plant operations. Since the mid-1990s the CAP has been concerned about emissions, so the company responded with a strategy of continuous improvement and integrated a business case for achieving emissions reductions, which is a key aspect of industrial ecology [22], the multi-disciplinary study of industrial material and energy flows with a socio-environmental consideration. In 2002, the company was able to find a market for sodium bisulfite (SBS) and developed a process to convert sulfur dioxide into this product, therefore greatly reducing emissions. There is a market for SBS in the pulp and paper and municipal raw water and waste water treatment industries, so this by-product has become a financial benefit for the company, while also providing an environmental benefit. In 2009, an acid plant scrubber was added to the process, again at the request of the CAP and reduced emissions even further. Sulco's emissions data demonstrates that the installation of the acid plant scrubber helped Sulco significantly reduce their sulfur dioxide emissions [1]. Upon examining Fig. 3.2, it is clear that these milestone investments resulted in major decreases in sulfur dioxide emissions.

Internalizing the values of RC does not always mean that community requests are answered immediately or at any cost, rather they are taken into consideration and acted upon in the context of continuous improvement. While this may not be in the best interest of any one stakeholder, it strikes a balance between the needs of all stakeholders in an open and transparent manner [23]:

> "When a problem is identified, they [Sulco] create a range of possible solutions and examine each option against environmental, financial and social objectives before determining a solution. Sometimes the optimal solution is not financially feasible in one calendar year and the company is good at bringing the preferred option forward along with the sense that it has a three-year time-frame, and asking the community (through the CAP) if they are supportive of waiting three years for an optimal solution, or implementing a sub-optimal solution immediately. This demonstrates a good grasp of the primary directive of RC which is to optimise, within obvious practical limitations."

Overall, while it is difficult to attribute which process innovation was motivated by an internalization of industrial ecologic principles versus the principles and codes of RC, it is clear that RC has had a major impact on the plant staff, members of the CAP, and on the reduction of emissions at the facility. One could also argue that RC incorporates the concepts of sustainability and industrial ecology, and therefore, a company who internalizes these principles, in turn, is a RC company (see Chapter 1).

3.3.2 Responsible Care and its influence on a corporate culture of sustainability at Sulco

The first independent, third-party RC verification takes place when a company states that their performance meets an expected level, which is the RC-In-Place. The verification then confirms that the company is adhering to the principles of RC, practicing RC, and working towards continuous improvement. Verifications are done every three years after formal acceptance of the first verification to ensure RC is rooted in all of a company's operations [21]. RC verification teams are generally comprised of CAP members, residents, former chemical industry employees, consultants, environmentalists, and academics. The verifiers are not paid by the company being verified and have no financial stake in their operations. This is instrumental to the development of a robust, independently verified program [21].

Sulco's "In-Place" verification took place in October 1993, as one of the first CCPA members to volunteer [24]. In 1996, an RC-In-Place verification was attained by Sulco. Since then, five verifications have taken place between 2000 and 2013. Generally, the verification teams have found that the RC ethic is well established and understood within the company and provides guidance to management and employees. The 2000 verification states that it is evident the company spent a lot of time and money to comply with the RC-In-Place report findings [24]. The 2003 report found that Sulco has strong management systems and is committed to both the CCPA's and Canadian Association of Chemical Distributors' (CACD's) versions of RC and Responsible Distribution [21]. The company has an overall management system that incorporates ISO 9000-2000, ISO 14000, RC, and Responsible Distribution. The RC management system at CCC/Sulco is "Embraced by senior employees and this, along with senior management commitment, will likely result in its sustainability" [25].

The 2000 report identified that Sulco's design of the oleum tank and rail cars were considered an industry best practice. The verification team also noted that overseas shipments are made only on "conference lines", which are marine shippers that adhere to safe transportation standards [24]. The report determined that the plant's CAP had been effective for ongoing dialogue with commercial neighbors and other representatives, but that the CAP should include more residential neighbors to improve the process. The report noted that CAP members on the re-verification team were very satisfied with the company's efforts to follow through on CAP recommendations, especially in the construction of buildings to house oleum storage tanks [24]. The report identified some areas of improvement including: working on worst-case scenario and emergency knowledge level of neighbors, installation of sulfur dioxide warning devices, broader residential neighbor involvement in the CAP, development of a management system for industrial hygiene, and development of a process to test the plant's emergency plan annually with the community.

The 2003 verification report highlighted some of the best practices implemented by Sulco since the previous verification. These best practices demonstrate that RC helps the company stay on track to work on findings from previous verifications and to continuously work towards improvement and developing best practices. Some of the best practices that Sulco implemented were: containment of oleum tanks for spill prevention, greenhouse gas reduction through recuperating by-product heat and selling energy to another company, development of an environmental, health and safety scorecard to monitor their own progress toward objectives and goals, installation of a computerized tank level monitoring system for spill and leak prevention, installation of pressure/vacuum relief vents on CCC solvent tanks at the Brampton facility to reduce vapor escaping from tank breathing, and continuous sulfur dioxide emission monitors on equipment to control the process to stay below sulfur dioxide regulatory limits [21]. The 2003 report discusses the community concerns in Elmira surrounding the chemical industry since environmental issues in the area forced the community to abandon their own water supply. Despite community concerns, Sulco CAP members (interviewed by the verification team) commended Sulco for "Their responsible chemical operation, their communication efforts and their concerns for the community in which they operate" [21].

Sulco has a very positive relationship with its CAP and its members have praised them for being leaders in community consultation [25]. An extra mile effort noted in the report is that the plant reduced the impact of the oleum worst-case scenario through recommendations from the CAP. The recommended option of building an enclosure for the oleum tanks was a higher cost than other options, but was the solution most acceptable to the community. The verification team found that improvements can be made in terms of increasing the frequency of CAP meetings and developing a method to discuss issues at CAP with the larger community [25]. Another opportunity for improvement identified by the team was for Sulco to determine what materials are being carried in tank cars past the plant, so they can identify risks and develop a plan for a rail emergency. CCC, Sulco's parent company, has worked to minimize process risks across its entire business [25]. The 2006 verification found that "Sulco's 5-year National Emissions Reduction Masterplan (NERM) is tied into projects that result in actual emission reductions" [25]. The company met their annual sulfur dioxide reduction targets for 2005 and 2006 and they reduced their water use.

It is obvious upon reviewing the third party verifications that RC codes and practices have benefited the company and the community both environmentally and financially. The analysis demonstrates knowledge and application of RC at all levels of the company from the parent company, CCC, Board members, management, and staff. However, one potential flaw of RC is that the company has the choice to implement whatever they wish, despite the recommendations of the verification. If an RC company is not verified and does nothing about it, it is still a part of the RC program and the CIAC. However, these companies will still have to be held accountable to their peers and the CIAC. In Sulco's case, RC has helped the company improve because the company genuinely embraces it, and is not using RC as greenwashing, which involves using exaggerated or falsified facts to make a company seem sustainable to its customers and the public.

Sulco plans for the worst-case scenario and ensures that the plant and community are prepared, as part of their commitment to implementing RC principles; in this case the Community Awareness and Emergency Response (CAER) principle of RC. As the verification reports describe, in order to minimize exposure the company focused on two areas: oleum-tank safety and rail-car safety. Road transport was also very important at the time. Through the RC worst-case scenario planning, oleum-tank leak was determined as a worst-case scenario that would impact the community negatively and pose a health risk. In order to minimize the effect of oleum-fume seepage, a half wall was proposed as a measure that was over and above regulation. However, the CAP, which was heavily involved in the design process, asked for a full wall and enclosure to be built to completely eliminate exposure to the community [25]. The company complied and reviewed the situation, examined the risk benefit and cost benefit of the option, and decided to build the more expensive enclosure. To minimize issues with transportation, the company invested in double-thick-wall rail and truck carriers with temperature maintenance, which are much more than regulation dictated at the time. While these are now the standard for oleum shipments greater than 30% strength, the company was forward thinking and met a future standard on their own timelines and has applied this improved design for oleum shipments of 20% and 25% oleum. The company also invested in other transportation-related measures, in collaboration with customers, including route changes in order to avoid exposure in the event of an accident and avoiding the morning and afternoon commuter rush as well as extreme weather events. In addition, the company increased their reliance on rail and decreased their reliance on truck carriers. These measures improved air quality and greenhouse gas emissions, as well as safety [18].

When asked whether their certification under ISO 14000 would have produced the same result as their participation in the RC process, Ron Koniuch, Vice President [18] agreed that to some extent ISO 14000 would have led to some containment, but many of the other aspects of the project would not be mandated:

> "RC worst-case scenario evaluation reduced the potential exposure of the community to an oleum release. This analysis is not regulated, however, you could argue ISO 14000 may have eventually dictated a reduction in risk. What would not have been part of 14000 is the presentation with the CAP members which did change the design of the building. On the SO_2 emission reductions, this was also not regulated and became a primary target on our ISO 14000 list due to the community involvement process. It may not have been regarded as high a priority without this communication."

The 2009 verification report continues to identify the fact that the RC ethic is well established in the company: "From the inception of RC the senior executives and various managers of the company have been very active in the promotion of RC inside and outside the company" [26]. The company has its own audit process for annual review of RC codes. The team found that the RC management system is "self-healing" and that the RC ethic is well understood and guides company actions and decisions. The basis for the company's environmental management system is the ISO 14001:2004 standard, which the plant is certified to. One of the significant changes noted since the previous re-verification was that an acid emissions scrubber was installed to reduce sulfur dioxide emissions. The verification team judges the following as industry best practices: Elmira plant's greater than 90% sulfur dioxide reduction, requirement for minimum fuel efficiencies on leased corporate use passenger vehicles, and the initiation of a process to measure the company's carbon footprint [26].

3.4 Analysis

3.4.1 Quantifying the impacts of not implementing Responsible Care at Sulco

This analysis has demonstrated Sulco's positive implementation of RC and that its impact on plant operations is significant, but it remains unclear as to the extent that

Sulco's successes can be attributed to RC. In order to evaluate the effectiveness of RC, it is important to investigate the cost of not participating in RC. As discussed, some of the major investments Sulco has made to reduce emissions and improve safety include the addition of the SBS plant, the installation of the acid plant scrubber, and the worst-case scenario planning which resulted in the installation of the oleum-tank building and improvements to transportation safety.

As previously discussed, in 2000 Sulco developed an environmental plan to reduce emissions, improve efficiency, and achieve cost savings, to ultimately increase production without increasing emissions. During this time, a SBS plant was developed in order to produce the by-product for the pulp and paper and municipal raw water and waste water treatment industries. In 2009 there was an expansion to the SBS production, in which a scrubber on the acid plant was added that is used to reduce sulfur dioxide emissions and compliment the main SBS production process. The plant focused on energy reuse and reduced greenhouse gas emissions by recuperating by-product heat from one of their processes and selling the energy to another company for use in their process. Sulco also increased their reliance on rail and maintained their reliance on truck carriers through the plant expansions, which helps decrease greenhouse gas emissions and improve air quality.

Using the sulfur dioxide (SO_2) emissions and the sulfur (S) tonnage, which was provided by R. Koniuch, Sulco Vice President, as an indicator of production, from 1990–2011 (Sulco data source), an analysis was conducted to demonstrate the impact of RC. From 1990–1994, the SO_2 emissions ranged from 0.052 to 0.078 tonnes SO_2 emitted/ tonnes S used. From 1995–2008, the emissions ranged from 0.022 to 0.030 tonnes SO_2/tonnes S. These emission reductions can be explained by the company's environmental plan and installation of the SBS plant. It was only in 2009 after the installation of the scrubber that the emissions significantly changed to 0.004 tonnes SO_2/tonnes S. In 2010 and 2011, the emissions were 0.002 tonnes SO_2/tonnes S. If the scrubber was not installed in 2009, the tonnes SO_2 emitted per tonnes S used could have remained at the 2008 level (0.022), which had been relatively constant since 2004. In 2011, the highest amount of sulfur was used in the process (17,499 tonnes). Without the scrubber installation, the 2011 emissions would have been comparable to 1995 levels (300–400 tonnes of SO_2). From 1990 to 2011, Sulco has reduced emissions by 96% (870 tonnes SO_2 in 1990 vs. 30 tonnes SO_2 in 2011). Without the installation of the scrubber, Sulco would have only reduced emissions by 55% between 1990 and 2011.

The result of the RC worst-case scenario planning to ensure that the plant and the community are prepared for emergency situations was an advanced oleum-seepage containment system. Originally, to reduce the risk of oleum seepage, a half wall was proposed as a measure that was over and above regulation. It was through the CAP process that a full wall and enclosure of the oleum tank was proposed in order to completely eliminate exposure to the community. Although this solution was more expensive, Sulco built the enclosure. Without RC, Sulco may not have a CAP or may have one that is less effective. The CAP is important to the RC Community Aware-ness Program for sites that warrant an advisory panel and regular RC verifications ensure that Sulco maintains a good relationship with the CAP as well as ensure that they follow through on their plans. Similarly, RC and the CAP helped Sulco improve their transportation safety through rail, truck, and routing improvements as well as more robust corporate policies. By anticipating future regulations, the transportation systems improvements were instituted on Sulco's plan and timeline rather than on the timeline imposed by regulation. This gave Sulco a competitive advantage. Sulco does not necessarily have a financial advantage over the other companies in the area, but they have a reputational advantage and have built trust in the community. This allows them to focus on building more support rather than avoiding scrutiny and dist-rust amongst the public and regulators.

It is clear that without RC, Sulco risked poor community relations, reduced worst-case scenario planning, increased emissions, and a less competitive business model. However, without RC, the company would still have a sustainability plan, innovative solutions to reduce waste and increase revenues, as well as some emission reduc-tions. The benefits of RC outweigh the costs of implementation, especially where RC had a direct financial benefit. Furthermore, as was demonstrated throughout the ana-lysis of verification reports for Sulco, positive community engagement, while hard to quantify in financial benefits, has clearly distinguished Sulco from other chemical plants in the area and provided a social-ecological benefit. This is consistent with stakeholder theory [8, 27], which emphasizes the multi-stakeholder approach and the engagement of non-financial stakeholders as a key to a firm's enhanced performance. Finally, although more funds were spent on additional, non-regulated construction and processes encouraged by the CAP, the company remains profitable and meets shareholder expectations [18].

3.4.2 Benefits of Responsible Care as it relates to industrial ecology, life cycle assessment, and resiliency

RC has three main components: a statement of the RC ethic and principles – which must be visibly tangible to internal and external stakeholders, management systems – through which continual improvement can be delivered in all aspects of RC, and the detailed Code elements that describe specific issues and practices that companies must consider in the areas of operations, accountability and stewardship. Through the ethic and Codes of Practice, RC companies are encouraged to: exceed laws, perform life cycle analysis on products, seek and address public concerns, use a pre-cautionary approach to chemical manufacture, understand that all stakeholders have the right to be informed, lead in the public policy process, provide mutual aid and peer pressure, and seek advocacy input and integrate the codes and practices into corporate decision making [28].

Many of the key aspects of RC relate to industrial ecological principles and sustainable development principles and have been effectively applied to chemical production and decreasing the release of toxic chemicals. However, Wernet et al. [22], found that this only amounts to 10% of the environmental impact of the chemical industry, while 50% or more of the industry's impact comes in the form of energy use. Energy use has impacts related to climate change and indirect emissions such as steam. While RC has been tracking plant energy use since 2005, this has not been the focus of much action [22]. However, as an early adopter, Sulco has found ways to increase returns through optimization of the internal usage of steam to reduce electrical or natural gas consumption and the sale of steam energy to a neighboring plant; one way that they decrease indirect emissions, while increasing returns. Parts of the steam-generation project were installed in 1983, prior to the implementation of the RC codes. This is consistent with the "waste equals food" aspect of the "cradle-to-cradle" product stewardship concept [29]. The related concept of resiliency: the ability of a system to absorb disturbance and still retain its basic function and structure [30], takes a system's approach to managing resources and industrial inputs and outputs. This includes the ability of a product to be reused or be biologically or technically metabolized. Rather than a sole focus of optimizing and increasing efficiency of certain aspects of a process, these research-practitioners advocate for increasing the efficiency of the entire system, including the impact on the ecosystem at the end of life. When the system as a whole is not taken into account, negative environmental impacts can increase, even though a component process of the plant is being made more efficient.

McDonough and Braungart [29] and Walker and Salt [30], cite numerous examples as to how these sustainability concepts have many financial, environmental, and social benefits. In one example, a chemical company was working to create an environmentally neutral fabric that filtered out potentially harmful wastes at the end of a product's life cycle in the product's design phase. This process was able to remove the need for harmful additives, thereby making the fabrics and solvents easier to manufacture and more cost effective. Furthermore, the effluent of the process was as clean as the influent, making it possible to use the effluent as an input, further lowering costs. This has important implications for RC. If traditional products that have harmful wastes or cannot be repurposed at the end of their useful life can, instead, be designed to have no impact or a positive environmental impact, they will exceed regulations to the point where no regulation is required [29].

As RC continues to evolve, it can build on concepts such as inherently safer designs that the cradle-to-cradle, sustainability, and resiliency schools of thought advocate for. If companies such as Sulco can internalize industrial ecology as a guiding principle, they can extend this to gain a competitive advantage with a more optimal process that is focused on total system sustainability. This will allow RC to have a greater impact in encouraging process innovation, which can lead to improved public relations, community support, and a reduced need for government inspection and oversight.

3.5 Conclusion

The goal of this study is to determine whether Sulco's environmental, social, and community engagement successes could have been achieved without the use of RC. The central question of the study asks, *"Does implementing Responsible Care contribute to better environmental, economic and social outcomes?"* This analysis focused on stakeholder interviews, production data, and RC verification reports to determine the relationship between CSR, resiliency thinking, industrial ecology, sustainability, and RC. The analysis demonstrated that a self-regulatory set of principles, policies, and programs that a corporation adopts can be effective in producing a positive impact on all stakeholders, including the community, environment, employees, and residents, if the company internalizes the values of CSR and its complementary concepts.

The Sulco case demonstrates that when production data, community relations, and comparison with companies that do not internalize RC and CSR to their fullest extent are taken into account, it is evident that a part of Sulco's success can be directly attributable to their participation in the RC program and their subsequent positive performance on re-verifications. Without RC, it is likely that Sulco would be compliant and a community member in good standing; however, their social, environmental, and financial success would not have been as effective and their reputation would not have been as positive. The benefits of Sulco's verified RC program allow the company to be ahead of regulations, implement efficiency programs at their own pace, and remain profitable by planning their plant improvements in phases; profiting from industrial ecological programs to reduce wastes and virgin material inputs.

Sulco's RC program is proof that the program provides real community value and can help assure residents that their health, safety, and interests are considered as part of regular plant operations. It should encourage residents and municipalities to apply pressure to non-RC companies to become verified and demonstrate positive results through the re-verification process. However, the community and observers need to also inspect verifications to ensure the company is self-healing and not using RC as a greenwashing tool. The Canadian Chemical Industry Association (CCIA) may need to implement more significant sanctions for companies that repeatedly have issues with verifications and better publicize these results.

As stated from the outset of the study, there are national policy implications for this research, as there is a case to be made for encouraging or requiring large emitters to have effective industry-led management programs that incorporate third party verifications and effective sanctions. While there is still room for improvement and growth, this could be an effective way to further improve air quality, reduce greenhouse gas emissions, and eliminate other environmental impacts, without large increased costs to government regulators.

References

[1] National Pollutant Release Inventory - Historical Substance Reports for Sulco Chemicals – Elmira Plant. Environment Canada, 2012. (Accessed July 10, 2012, at http://www.ec.gc.ca/ pdb/websol/querysite/facility_history_e.cfm?opt_npri_id#equal#0000001156&opt_report_ year#equal#2011.)

[2] Clausi, R. Personal communication. June 2012 (details available on request from the authors).

[3] Elmira, Ontario (Code 0258) and Ontario (Code 35) (table). Census Profile. 2011 Census. Statistics Canada Catalogue no. 98-316-XWE. Statistics Canada, Ottawa, 2012. (Accessed July 10, 2012, at http://www12.statcan.gc.ca/census-recensement/2011/dp-pd/prof/index. cfm?Lang#equal#E.)

[4] Control Order: Uniroyal Chemical (ONTR26/26R950.51). Ministry of the Environment, Toronto, 1999. (Fax available; reference code ONTR26/26R950.51).

[5] Bélanger, J. M. Responsible care in Canada: The evolution of an ethic and a commitment. Chemistry International 2005;27(2):1.

[6] Two decades of responsible care: Credible response or comfort blanket? Environmental Data Services (ENDS) Ltd., No. 360, London, 2005. (Accessed March 25, 2007, at http:// www.responsiblecare.org/filebank/ENDSFeaturesResp_Care.pdf.)

[7] Lenox, M. J., Nash, J. Industry self-regulation and adverse selection: A comparison across four trade association programs. Business Strategy and the Environment 2003;12(6):343–356. (Accessed July 10, 2012, at http://www3.interscience.wiley.com/cgi-bin/fulltext/106560527/ PDFSTART.)

[8] Niskanen, T. A Finnish study of self-regulation discourses in the chemical industry's Responsible Care programme. Business Ethics 2012;21(1):77–99.

[9] Moffet, J., Bregha, F., Middelkoop, M. J. Responsible care: A case study of a voluntary environmental initiative. In: Webb, K., ed. Voluntary codes: Private governance, the public interest and innovation. Carleton Research Unit for Innovation, Science & Environment, Ottawa, 2004;177–208.

[10] Prakash, A. Responsible care: An assessment. Business & Society 2000;39(2):183–209.

[11] Steelman, T. A., Rivera, J. Voluntary environmental programs in the united states: Whose interests are served? Organization & Environment 2006;19(4):505–526.

[12] Tschopp, D. J. Corporate social responsibility: A comparison between the united states and the European Union. Corporate Social Responsibility and Environment Management 2005;12: 55–59.

[13] Givel, M. Motivation of chemical industry social responsibility through responsible care. Health Policy 2007;81(1):85–92.

[14] King, A. A., Lenox, M. J. Industry self-regulation without sanctions: The chemical industry's responsible care program. Academy of Management Journal 2000;43(4):698–716.

[15] Reisch, M. Responsible care: The chemical industry has changed its ways, but more radical actions may be needed to change public opinion. CENEAR 2000;78(36):21–26. (Accessed July 10, 2012, at http://pubs.acs.org/cen/coverstory/7836/7836bus1.html.)

[16] Topalovic, P., Krantzberg, G. Broken Link: With supply chains strung across the developing world how can the chemical industry ensure that the end product is safe? Canadian Chemical News 2010;62(7):21–25.

[17] West, B. Personal communication. August 1, 2012 (details available on request from the authors).

[18] Koniuch, R. Personal communication. July 10, 2012 (details available on request from the authors).

[19] Kelly, K. Personal communication. May 2013 (details available on request from the authors).

[20] Bélanger, J. M. Personal communication. June 2012 (details available on request from the authors).

[21] Blondin, R. Responsible Care Re-verification of Canada Colors and Chemicals Limited and Sulco Chemicals Limited. 2004. (Accessed July 10, 2012, at http://www.canadianchemistry.ca/ResponsibleCareHome/VerificationBR/tabid/103/ItemId/70/vw/1/Default.aspx.)

[22] Wernet, G., Mutel, C., Hellweg, S., Hungerbuhler, K. The Environmental Importance of Energy Use in Chemical Production. Journal of Industrial Ecology, 2011;15(1):96–107.

[23] Vincett, J. Personal communication. June 2012 (details available on request from the authors).

[24] Dillabough, C. D. Re-verification For Sulco Chemicals Limited and Canada Colors and Chemicals Limited. 2000. (Accessed July 10, 2012, at http://www.canadianchemistry.ca/ResponsibleCareHome/VerificationBR/tabid/103/ItemId/100/vw/1/Default.aspx.)

[25] Whitcombe, G. Responsible Care Re-verification 2005–2008 of Canada Colors and Chemicals Limited and Sulco Chemicals Limited. 2007. (Accessed July 10, 2012, at http://www.canadianchemistry.ca/ResponsibleCareHome/VerificationBR/tabid/103/ItemId/63/vw/1/Default.aspx.)

[26] Dillabough, C. D. Responsible Care Re-verification Sulco Chemicals Limited and Canada Colors and Chemicals Limited. 2009. (Accessed July 10, 2012, at http://www.canadianchemistry.ca/ResponsibleCareHome/VerificationBR/tabid/103/ItemId/131/vw/1/Default.aspx.)

[27] Freeman, R. E. Strategic Management: A stakeholder approach. Pitman, Boston, 1984.

[28] Responsible Care Codes of Practice. Chemical Industry Association of Canada, Toronto, 2008.

[29] McDonough, W., Braungart, M. Cradle to Cradle: Remaking the Way We Make Things. North Point Press, New York, 2002.

[30] Walker, B.H., Salt, D. Resilience Thinking: Sustaining Ecosystems and People in a Changing World. Island Press, Washington, D.C., 2006.

Peter Topalovic and Gail Krantzberg

4 Responsible Care workshop: Exploring the effectiveness of Responsible Care – workshop overview and toolkit

Abstract: This workbook provides a workshop facilitator with the necessary tools to conduct a Responsible Care workshop in other organizations with ease (Appendix A contains a list of the necessary files that accompany this workbook). The workshop explores the effectiveness of Responsible Care using the results of a case study originating in Haiti and involving poor supply-chain management, chemical contamination, and negligence. The workbook also summarizes the results of a pilot workshop conducted at the McMaster University Centre for Engineering and Public Policy (Hamilton, Ontario Canada) in partnership with the International Union of Pure and Applied Chemists (IUPAC) on May 13, 2010.

4.1 Workshop agenda

A full day workshop is organized in the following manner:
Pre-workshop assigned readings:

- Responsible Care: History and Development.
- Haitian Cough Syrup Contamination Case Study, with analysis of Responsible Care factors hidden from participants.
- Others, as selected by the organizers.

Schedule:
8:30 am: Refreshments, Registration and Introductions
9:00 am: Overview of the day by workshop facilitator
9:10 am: Presentation 1 – Plenary Review of Responsible Care and Workshop Purpose
10:00 am: Presentation 2 – Haitian Case Study Introduction
10:45 am: Break
11:00 am: Breakout Sessions Overview

- Purpose of breakouts explained and questions reviewed.
- Three breakout discussion groups formed.

11:10 am: Breakout Group Discussions
12:00 pm: Working Lunch, Breakout groups develop formal position
1:30 pm: Group 1: Question Reviewed and Discussion of Results

1:45 pm: Group 2: Question Reviewed and Discussion of Results
2:00 pm: Group 3: Question Reviewed and Discussion of Results
2:15 pm: Case Study Results Reviewed
– Revealing the actual results of the case study and how they differed from the group's expectations.
– Analysis from Workshop Facilitator.

4:00 pm: Closing Remarks and Adjournment

4.2 Workshop logistics

The workshop is conducted for trainees with little prior knowledge of the Responsible Care (RC) ethic or the case study reviewed in the workshop. Trainees should be organized into groups of 10 at a maximum. One facilitator should be provided for each group who has extensive knowledge of RC. The following presentations can be made available to the workshop organizer by contacting the authors of this book, the CIAC, or the IUPAC.

Presentation 1 explains the RC ethic and how it differs from traditional environmental management systems in order to ensure all participants understand the basis of what was explored in the case study. The presentation material summaries are provided in Section 4.3 and Section 4.4 of this document and the full readings are provided in Appendix A.

Presentation 2 provides an overview of the case study (the presentation is provided in Appendix A) and sets the stage for the discussion questions. With the context of the problem set, the questions are explained, groups are formed, and the discussion begins. Facilitators answer questions and foster discussion. The purpose of the discussion questions is to help the workshop attendees analyse the case study and the role of RC to determine if "Responsible Care, appropriately applied throughout the supply chain, would have averted the crisis experienced in Haiti and other developing countries?" They are able to form their own opinions and complete their own analysis of the case study. Participants then move back into plenary to present their findings on the theory assigned.

The three theories that the groups breakout to discuss are the three answers to the central question of the case study which is outlined in Section 4.5.

Once the discussion and analysis has occurred, the full results of the study are revealed and discussed, and the workshop outcomes and real-life outcomes are compared and contrasted. Overall the discussion should lead to the fact that RC, appropriately applied throughout the supply chain, would have averted the crisis experienced in Haiti.

In preparation for this workshop, attendees are sent a package that includes a document on the history and development RC and a Case Study overview (see Chapters 1 and 2).

4.3 Responsible Care overview

RC is an environmental management system which compliments regulations and drives a company's performance beyond the minimal environmental standard. It was developed in the 1980s to meet the threat of costly new regulations and a decreasing public trust in the chemical industry. It consists of an ethical set of principles as well as management system of codes and practices which aim to address environmental, ethical and social concerns of chemical manufacture over their entire life cycle. Throughout this chapter, the use of the term RC will refer to the Canadian implementation as outlined by the Canadian Chemical Producers' Association (CCPA) in the "Ethic and Codes of Practice of Responsible Care". This includes elements such as: external third-party verification of compliance, the combination of ethical and prescriptive measures in the implementation of the code, the formation of Community Advisory Panels (CAPs), the use of an independent National Advisory Panel (NAP), the demonstration of RC in all areas and levels of the corporation, and the requirement that each member company implements all the RC codes of practice to maintain their membership in the industry association.

Key characteristics of RC to focus on as its development and future are discussed include:
- Responsible Care is a *process* not a *project*, meaning that it continues to evolve as the participants listen and react to growing and changing needs.
- The needs of the industry arise from issues surrounding the environment, social responsibility, and economic sustainability.
- One of the main evolutionary components of Responsible Care has been the "rebranding" of the role(s) of chemistry and the chemical industry.
- Chemistry-based industry now covers many new subdivisions including nanotechnology, biotech industry, materials research, and more.
- Chemistry-based industry also involves the supply chains from resource extraction to consumer use to waste and recycling of materials.

The Principles of Responsible Care include the statement: "We are committed to do the right thing, and be seen to do the right thing. We dedicate ourselves, our technology and our business practices to *sustainability – the betterment of society, the environment and the economy*. The Principles of Responsible Care are the key to our business success, and compel us to:
- Work for the improvement of people's lives and the environment, while striving to do no harm.
- Be accountable and responsive to the public, especially our local communities, who have the right to understand the risks and benefits of what we do.
- Take preventative action to protect health and the environment.
- Innovate for safer products and processes that conserve resources and provide enhanced value.

– Engage with our business partners to ensure the stewardship and security of our products, services and raw materials throughout their life cycles.
– Understand and meet expectations for social responsibility.
– Work with all stakeholders for public policies and standards that enhance sustainability, act to advance legal requirements and meet or exceed their letter and spirit.
– Promote awareness of Responsible Care®, and inspire others to commit to these principles."

RC differs from other environmental management systems according to Table 4.1 and this should be emphasized in the workshop.

It should be noted that no company or corporation is created equal. Laggard companies who do not follow the principles of RC may exhibit some or all of the attributes found in the left column. Some environmental management systems allow these activities and attitudes to prevail, while still granting the companies certification. These companies could also exhibit some of the tendencies found in the right side column; however, RC works to guarantee these elements are included in a company's ethic.

One important element of the RC ethic is the built-in requirement of growth and analysis through company and community feedback and the oversight of the independent NAP. This growth can be characterized by examining the ethic as it existed in the early 2000s and the ethic today, which has become more progressive and integrates the new ideas and philosophies of industrial ecology and sustainability (Table 4.2).

The language used in the new version of the principles makes explicit some items that were implied in the past and also holds companies more responsible for their business decisions and actions.

Table 4.1: Comparison of other environmental management system and Responsible Care

Other strategies and non-complying companies	Responsible Care companies
✗ meet the law	✓ do right thing (exceed laws)
✗ resist new laws	✓ work for strong regulations
✗ keep a low profile	✓ be seen to do the right (or wrong) thing
✗ product liability – customers' problem	✓ life cycle stewardship – value-chain
✗ downplay public concerns	✓ seek & address public concerns
✗ manage risk	✓ inherently safer products and processes
✗ do less "bad"	✓ improve people's lives and environment
✗ chemicals innocent until proven guilty	✓ precautionary approach
✗ continuous improvement	✓ innovation
✗ hazard information to who needs it	✓ public, employees right to understand
✗ every company for themselves	✓ mutual aid & peer pressure
✗ ignore or fight activists	✓ seek activists' input
✗ do not require a role in "social responsibility"	✓ understand and meet social responsibility
✗ bottom line & laws guide decisions	✓ integrate all above into decisions

Table 4.2: The evolution of the Responsible Care ethic

Early 2000s	2008 – present
sustainability implied	sustainability "hard wired"
do less harm	improve lives & environment
reduce risk	safer products and processes
continuous improvement	innovation
"precaution" implied	"take preventative action"
conservation implied	conserve resources
"security" not in Principles	"stewardship and security"
"we are stewards ..."	stewardship with partners
"respect all people"	social responsibility
improve, but resist, laws	advocate for sustainability laws
inspire others to commit	branding of Responsible Care

4.4 Case study overview

In 1996, 80 Haitian children died from ingesting cough syrup tainted with diethylene glycol (DEG), a chemical commonly found in antifreeze. An investigation conducted by the United States Food and Drug Administration (FDA) found that Pharval, a local company who produced the cough syrup products, Afebril and Valodon, did not contaminate the product at their site. Instead, a supposed pharmaceutical-grade shipment of glycerin, a key component in the most widely prescribed cough syrup in the country, was contaminated at its source in China. However, the Haitian company was under the assumption that the chemical was produced in Germany from chemical giant Helm AG, parent of VOS B.V. and, as such, felt that it did not have to implement any quality controls on the imported European product.

Issues surrounding product stewardship, supply-chain management, quality control, politics, and legal responsibility are at the heart of this disaster and combine together to create a complicated web of interactions leading to negative implications for all parties involved. In the aftermath of this disaster many of the companies in the supply chain engaged in finger pointing, lawsuits, and denial of responsibility.

This case study addresses questions that arise from this disaster in the context of RC. We test the hypothesis that the magnitude of this crisis would have been reduced if some or all of the companies along the supply chain subscribed to RC principles and codes. This will serve to evaluate the strengths and short comings of RC.

4.4.1 Accountability for poor management of the supply chain and a lack of quality control

The United States FDA's investigation of the cough syrup incident indicated that the pharmaceutical manufacturing and testing facilities at Pharval laboratories did not meet international standards. This was due to a variety of factors, including the fact that Haitian regulations are not as strict as those in other countries. Furthermore, the maintenance costs for clean rooms, proper HVAC systems and high-tech testing equipment are too high for most companies in developing countries. According to Suzanne Synod, even if the facilities were clean and properly maintained, Pharval would not have access to the technologies required to test for DEG.

The investigation of the glycerin suppliers leads to some disheartening conclusions related to the management of international supply chains in China, the Netherlands, and Germany, especially where developing countries are concerned. In the Haitian case, no company along the supply chain was found to be directly responsible for the problems that occurred. Some denied responsibility, others covered up mistakes and in the case of the contamination source it was never determined which Chinese company was responsible; however, it is believed to be Sinochem, based in Peking.

The investigation of VOS B.V. by the Special Rapporteur of the United Nations Economic and Social Council of the Commission on Human Rights found that the company knew about the contaminated glycerin after it sent the shipment to Haiti but did not alert authorities. VOS B.V. sent a sample of the Chinese shipment to an independent testing laboratory and falsely marked the barrels of glycerin as 98 PCT USP, pharmaceutical grade. According to Kevin J. McGlue, a Board member of the International Pharmaceutical Excipients Council, "Where there is a loophole in the system, a frailty in the system, it's the ability of an unscrupulous distributor to take industrial or technical material and pass it off as pharmaceutical grade."

A certificate of pharmaceutical quality also accompanied the barrels, which was taken from the certificate that originated from China. It is common practice to re-use certificates when a product changes hands from manufacturer to supplier and onward to the destination customer. The identification of the source contamination and the FDA inspection of Pharval laboratories clearly established that the glycerin was contaminated at the source and not along the supply chain. It was a lack of product stewardship policies and a lack of due diligence across the supply chain that led to the disaster. In both Haiti and Panama, a more recent country which experienced DEG contamination, the factory's original certificate of analysis for the glycerin containers did not accompany them as they moved across the supply chain. Instead, a copy of the original was used and stamped with the receiving company's information each time the container exchanged hands.

Helm AG, one of the largest chemical companies in the world and the parent company of VOS B.V., declined to comment on the case, given that the contamination occurred outside of Germany. Helm AG has been associated with other issues involving the transport of materials to the third world, according to the German media. The Chinese government also denied any responsibility since the glycerin was not shipped directly to Haiti from China. According to Ms. Pendergast, a private lawyer and consultant, China has the most to answer for. "Everybody else is just reacting to initial failures … It needs to take steps to protect not just its own consumers but also consumers all around the world".

No international supply-chain management regulations exist to solve problems such as this one; however the European Union's Registration, Evaluation, Authorization and Restriction of Chemical Substances (REACH) legislation and RC's supply-chain management policies may have a positive effect. This issue has received major coverage over the last ten years since the Haitian disaster occurred. In the Haitian case, there were many organizations responsible for transporting the glycerin across international boundaries and, therefore, it is very difficult to lay blame. However, after the details of the case were sorted out, some litigation proceedings were undertaken in the Netherlands, Germany, and Haiti.

4.4.2 Economic and social effects of the incident and the ensuing litigation

Pharval settled with the Haitian families whose children died from DEG exposure for $10 000 USD per family. They also filed a civil suit against VOS B.V. in the Netherlands, jointly with the families. The litigation against the companies involved focused on VOS B.V., since they knowingly sent DEG-contaminated glycerin to Haiti. Eventually a civil suit was settled outside of court with the Dutch company for the same amount as Pharval had settled with the parents. In the aftermath, the affected families were compensated, yet no company accepted full responsibility for the tragedy. VOS B.V. was also prosecuted by the Dutch government, found guilty of the cover up, and fined $250 000 USD. Although both these companies were partly to blame, both Pharval and VOS B.V. (now Helm Chemicals B.V.) remain in business today.

Other attempted litigation has generally failed to produce a favorable outcome for the plaintiffs. The Chinese government and corporations would not work with the US FDA, who assisted Haitian officials with the investigation, to find the contamination source, denying responsibility and moving glycerin operations away from the site of contamination. David Mishael, a lawyer in the United States who has spent the last ten years since the Haitian disaster representing Haitian parents, has unsuccessfully pursued legal claims against Helm AG and VOS B.V.

In terms of the economic and social effects of the disaster and the ensuing litigation, very little change occurred and a minimal amount of punishment was received by all parties involved. The costs associated with non-compliance were $250 000 USD and very little transparency or accountability was demanded in the aftermath. Policies on supply-chain management remain largely unchanged world-wide, with the exception of the European Union's REACH legislation. Evidence of a policy and regulatory gap is clear in the past ten years of repeated DEG contamination in developing countries throughout the world. In many cases, the source of this contamination continues to be from poorly regulated Chinese suppliers.

Panama experienced a tragedy similar to that of Haiti, in 2006, with DEG being the contaminant in a government-manufactured cough syrup, resulting in hundreds of deaths. Over the years, other countries such as Bangladesh, Argentina, Nigeria, China, and India have also been affected by DEG-poisoning cases. The latest case in Panama clearly illustrates that the proper steps were not taken to minimize this preventable disaster. After the Haitian experience, an inexpensive DEG testing kit was developed to assist regulators in identifying contaminated shipments; however, this is not in widespread use.

DEG poisoning has now become a global problem with a preventable death toll in the thousands. China has recently taken action against drug counterfeiters, but when the role of pharmaceutical companies in the disaster in Panama was examined, it was found that no laws had been broken. Many developing countries, including China and India, need tougher regulations in order to meet safe standards.

After the Haitian incident, world health experts recommended improving the Certificate of Authenticity system to provide a clear path of the material flow through the supply chain from source to destination. It also stressed that transparency and accountability should be enforced through regulations and investigations within and between international borders. As long as counterfeiters do not fear prosecution, there is no incentive to improve the quality of their products.

4.4.3 Possible roles for Responsible Care

The issues surrounding global supply chains which originate and terminate in the developing world have continued to plague the chemical industry since first making news in Haiti and continuing to make news into the new millennium in countries such as Panama. In North America, issues with lead-poisoned toys, DEG-tainted toothpaste, and other dangerous products mimic the problems encountered in developing countries; however, in most situations, tainted products are identified before they reach the end customer. Tough North American regulations and industry standards, which are not usually present in developing countries, may be responsible for successfully identifying dangerous products.

This case study examines events in which DEG poisoning and drug counterfeiting have continued to be an issue in developing countries since the 1996 Haitian disaster. A mix of national and international regulations along with corporate voluntary initiatives is assessed as a way to determine whether the public can be better protected from these poisonings and tragedies. The central question of this case study asks:

Would Responsible Care, appropriately applied throughout the supply chain, have averted the crisis experienced in Haiti and other developing countries?

To answer this question and determine whether RC can have a positive impact, an investigation of global supply chains, due diligence, stakeholder trust and corporate culture follows. This case study analysis will examine evidence to determine whether or not the following theories are correct:

1. The companies involved did not internalize the concepts of product stewardship and the cradle-to-cradle philosophy that RC advocates.

2. The loss of business and reputation through erosion of trust was not a major consideration in the decision-making processes of the companies involved.
3. The major players involved in the case study did not embed the philosophies of the RC ethic or create a corporate culture of protection for all stakeholders.

The entire case study is included in Appendix A.

4.5 Workshop breakout sessions

The workshop breakouts are intended to explore the central question described in the case study: *Would Responsible Care, appropriately applied throughout the supply chain, have averted the crisis experienced in Haiti and other developing countries.* Each group discusses one anticipated finding and then presents their analysis to the workshop plenary. The analysis requires the group to answer the following questions:

–   Given the anticipated finding you have been assigned, do you think it answers the central question positively or negatively?
–   What evidence can you produce to back up your claim and strengthen your position?
–   Does any evidence exist that may go against your analysis?
–   What type of role does RC play in the future of the Canadian and worldwide chemical industry, given the results of your analysis?
–   Can similar RC-type, ethic-based systems be applied to other industries to complement existing supply-chain management and environmental management systems particular to those industries?
–   How important are government regulations and policies compared to industry-led, voluntary environmental management systems such as RC? Given the analysis presented in the case study, are regulations enough to regulate an industry? Could they be improved? Should voluntary ethics be relied upon, and if so, to what extent?

The following three anticipated findings to the central question form the different aspect of the case study that each group will discuss: *Would Responsible Care, appropriately applied throughout the supply chain, have averted the crisis experienced in Haiti and other developing countries?* This case study analysis will examine evidence to determine whether or not the following theories are correct:

1. **Group 1:** The companies involved did not internalize the concepts of product stewardship and the cradle-to-cradle philosophy that RC advocates.
2. **Group 2:** The loss of business and reputation through erosion of trust was not a major consideration in the decision-making processes of the companies involved.
3. **Group 3:** The major players involved in the case study did not embed the philosophies of the RC ethic or create a corporate culture of protection for all stakeholders.

The summarized results of the pilot workshop are included in Appendix B.

4.6 Pilot workshop analysis

These are the results of a pilot workshop conducted at the McMaster University Centre for Engineering and Public Policy (Hamilton, Ontario Canada) in partnership with IUPAC on May 13, 2010. This summary data is included in the toolkit to help the workshop facilitator guide discussion and anticipate questions and feedback from the groups.

The breakout sessions were highly engaging and covered the three anticipated findings well. Overall, the group analysis confirmed the results of the case study and contributed additional insight into the topic and the workshop as a whole. It is evident that different groups of workshop attendees from different industries and practical knowledge bases would produce slightly differing opinions on the topic and would therefore benefit from conducting this workshop according to the plan outlined in this document.

The general themes covered in the workshop and brought up in the brainstorming sessions included:
- Product stewardship
- Chain of custody
- Sustainable and green chemistry
- Lower environmental and societal footprint
- Inherently safer design
- Life cycle analysis
- Cradle-to-cradle design
- Supplier preference
- International standards, policies, and regulations
- Branding (strengthening and awareness)
- Provide the need for small customers
- Quality linked to Chief Executive Officer (CEO) compensation (versus profitability)
- Reputation (which could be location dependent)
- Broaden RC from producers in the industry association to other areas such as academic institutions, chemical designers, distribution and supply chains, other chemistry-based industry associations

In terms of moving RC towards future improvement and growth, the attendees suggested:
- Reduced insurance rates for RC members and other incentives are important.
- Tying performance to success is a key element that should be emphasized.
- Smaller companies could have less stringent requirements (i.e. level 1, 2, 3, max) so they are not overwhelmed with the costs of full compliance:
  - Potentially a phasing strategy to introduce RC gradually.
  - Work through the verification process and deal with issues as they arise.
  - Adapt codes to the company's needs.

- Issues with labor practices, especially in an international context, could be integrated into the code; currently there are no formal provisions or requirements.
- RC and life cycle analysis should be formally linked.
- UN and UNESCO Recognition for RC is required in developing countries.

It is important for the workshop facilitators to read and fully understand the information presented in the workshop toolkit (Appendix A), as it will enable the facilitators to compare and contrast the groups' analysis and outcomes with the actual events surrounding the Haiti case study. While the groups involved in the workshop conducted at McMaster University developed similar conclusions to the actual events, they also offered additional insights and in some instances disagreed with the findings of the paper. This is bound to happen in other instances of this workshop and therefore the workshop facilitator should be prepared to help the group understand the differences that exist and potentially develop actions for the Chemistry Industry Association of Canada to review.

4.6.1 Feedback from the Pilot workshop and breakout sessions

Positive Comments:
- Develops the concept of RC and introduces its use in industry.
- Highlights the distinction between regulations and ethical, culture-based, industry-led, voluntary environment management systems.
- Encourages good discussion regarding supply chain management.

Needs Improvement:
- Situation 1 and 3 are similar – make sure they are clearly separated.
- Separate problem definition from solutions discussion – help to focus on problem definition so that the discussion does not move into solutions discussion too quickly.
- Investigate other cases that occurred in North America, not just in developing countries:
  - Look at issues such as the listeria outbreak in 2009.
  - Help level the playing field in terms of the discussion.
- Give an overview of the codes or links to the RC website before the workshop for further review and background information.
- Present more than one case study that covers a variety of different issues the chemical industry faces:
  - Present at least one positive case to compliment the results arising from the negative case.
  - Highlight the success of RC.
  - Describe both the positive and negative cases.
  - Match the case study with the local area that the workshop is taking place – connect the workshop to local initiatives.

Overall, the feedback from workshop participants was overwhelmingly positive. They felt engaged, enlightened, and better educated on the topic. The connection to real-world events, systems, and solutions was highly valuable.

Acknowledgements

The development of this workshop toolkit was made possible with funding support from the International Union of Pure and Applied Chemists (IUPAC) and with the leadership of Jean Bélanger and Bernard West.

Appendix

A: Workshop toolkit

The resources listed here are found in this book.
Resources for Workshop Facilitator:
- Responsible Care: History and development (Chapter 1)
- Responsible Care in global supply chains: A case study (Chapter 2)

Resources for Workshop Attendees:
- Responsible Care: History and development (Chapter 1)
- Responsible Care in global supply chains: A case study (see Chapter 2), with the "Analysis" section removed, as this will be discussed in the workshop
- Responsible Care website – available online at the Chemistry Industry Association of Canada's website

Presentations:
The three files described here can be made available to the workshop organizer by contacting the authors of this book, the CIAC, or the IUPAC.
- Presentation 1: RC workshop – overview and context.ppt
- Presentation 2: RC workshop – case study overview.ppt
- Google Earth file: Haiti case study supply chain – Google Earth.kmz.

This file includes a fly over of the supply chain from country to country. Should this not work properly on the computer used for the presentation, the file contains a description of each country as follows:
- Haiti:
    - 80 children die from contaminated cough syrup under brand names: Afebril and Valodon.
    - Pharval, a company based in Haiti, is not responsible for the contamination.
    - FDA finds that laboratories are sub-standard.
    - Pharval obtains Glycerin from a German Company.
- Germany:
    - Pharval purchases the glycerin from a German company named Helm AG, a chemical giant.
    - Since the shipment came from Germany, Pharval does not test the glycerin.
    - Shipment is labeled "Made in Germany".
    - The container's certificate of pharmaceutical quality describes the shipment as 98% Pure USP.
- China – The Actual Source:
    - Actual point of origin for the Tainted Glycerin Containers.
    - Most likely the DEG was added to the glycerin because it is cheaper to produce.

- – Certificate of pharmaceutical quality is forged.
  - – Poor records, a lack of strict laws and regulations and little accountability result in an inability to accurately identify the company of origin.
- – The Netherlands:
  - – Vos B.V., wholly owned subsidiary of Helm AG buys the shipment from the unknown Chinese-based company.
  - – At the same time they send the shipment to Haiti, they send a sample for testing.
  - – Sample comes back as tainted, but no action is taken to notify Pharval.
  - – Certificate of pharmaceutical quality is copied from original containers.
  - – Would this have happened if the client was based in North America?

Room Requirements:
- – Lecture room with seating to accommodate all participants.
- – Whiteboard or chalkboard (for plenary discussion), laptop, projector.
- – Notepads (for breakout note taking).
- – Large poster paper (for plenary discussion summaries), markers.

B: Pilot workshop breakout sessions summary

Group 1 explored anticipated finding 1: *The companies involved did not internalize the concepts of product stewardship and the cradle-to-cradle philosophy that RC advocates.*

Group 1 found that there was no ownership of the product life cycle from the companies involved in the supply chain. The failures of each company along the supply chain were summarized to include key issues such as a lack of due diligence, use of discriminatory principles, lack of international standards and regulations, misuse of reputation and labeling, poor communication, and a lack of testing and analysis. The group reiterated that not only did the companies fail to internalize product stewardship principles; they acted illegally in some cases. They proposed possible solutions to improve product stewardship in the supply chain, including: international laws and regulations, policies and systems to ensure product safety and due diligence, auditing of product quality upstream and downstream, and to develop executive positions focused on product quality and developing RC principles.

See Appendix C for a full analysis.

Group 2 explored anticipated finding 2: *The loss of business and reputation through erosion of trust was not a major consideration in the decision-making processes of the companies involved.*

Overall, from the evidence, group 2 determined that the anticipated finding was accurate. They examined each company in the supply chain and cited evidence for their lack of concern regarding business reputation. The group determined that a company that did not act ethically would have been isolated by RC and, potentially, the act would have been avoided. This was considered particularly important as the supply chain grows and international regulations become complicated or ineffective. They also cited issues with investors and how important profit is over ethics.

See Appendix C for a full analysis.

Group 3 explored anticipated finding 3: *The major players involved did not embed the philosophies of the RC ethic or create a corporate culture of protection for all stakeholders.*

The group analyzed the finding in terms of the issues that arise in connection with embedding an ethic including transparency, accountability, preventative actions, risk assessment, innovation, stewardship, awareness and improvement; the codes of RC. The group found that the major issue with this finding was that the companies were not required to "do the right thing" because no legislation or market demand existed to ensure this occurred. It was determined that if the companies in the supply chain imbedded RC philosophies into their corporate culture, they would have been more likely to follow the RC codes and have avoided the disaster that occurred in the Haitian cough syrup case. They determined that initial costs, lack of branding and cultural differences impede the adoption and embedding of RC into corporate culture.

See Appendix C for a full analysis.

C: Pilot workshop breakout sessions raw data

Group 1: *The companies involved did not internalize the concepts of product steward-ship and the cradle-to-cradle philosophy that RC advocates.*

Our evaluation of the above statement in regards to the glycerin contamination in Haiti began with an examination of the RC failures that each company experienced and finishes with the recommendation of solutions and alternatives that each company could undertake in order to improve RC. In general, our evaluation found that there was no ownership of the product life cycle from any of the companies.

Table A.1 summarizes the RC failures of each company.

As demonstrated above, none of the companies involved in the supply chain gave proper attention to the upstream and downstream supply chain. Indeed, one of the companies (VOS B.V.) conducted itself illegally, in that it knowingly released a contaminated product and falsified records that showed otherwise.

Table A.1: Responsible Care failures at each company in the supply chain

Company	Responsible Care failures
Sinochem (China)	– Did not confirm purity of substance – Lack of standards (no laws were broken in this situation) – Have no concept of the entire supply chain and destination for their product (the downstream uses) – The need to remain competitive means sacrificing RC principles
VOS B.V. (The Netherlands)	– Found product to be contaminated but falsely labeled the barrels as 98% pure – Took no action to warn Pharval of contaminated product – Lack of understanding of the supply chain – Discriminatory principles with respect to notice of contamination – Lack of due diligence
HELM AG (Germany)	– Misuse of the "Made in Germany" logo serves as deception and falsification of the actual supply chain – Shipping to end user prior to testing and quality control measures – Failure to communicate with Chinese supplier to ensure quality of product – Discriminatory principles with respect to end user (lower standards in Haiti)
Pharval (Haiti)	– Lack of quality assurance, too much trust in supposed source of product (lack of understanding of supply chain) – Lack of testing of product – Extra effort required in the face of lax standards

Table A.2 summarizes our recommendations for solutions to improve RC for each company.

Group 2: *The loss of business and reputation through erosion of trust was not a major consideration in the decision-making process of the companies involved.*

Following the supply chain:
Chinese companies:
– Easily changing company names means reputation is not important.
– Far removed from the final product: not even identifiable.
– The falsifying of the certificate of authenticity shows no concern for accuracy.
– Not enough information on this portion of the supply chain to determine what could have been different if reputation had been a major consideration.

VOS B.V.:
– There could not have been a culture of caring about reputation, since the product was tested after being sent out, and the test results went nowhere.

Table A.2: Potential solutions to improve Responsible Care at each company

Company	Solutions/possible alternatives
Sinochem (China)	– Need a more thorough understanding of who they ship the product to and the likely end use of the product – Require a description of the overall supply chain from whomever they sell the product to, even if just to a major reseller (reseller must provide destination/end use) – ISO standards – Refuse to sell product if end user is not defined and complete supply chain is not understood – Engage government to develop quality regulations for the industry, so that the entire chemical industry is not affected by a few bad circumstances
VOS B.V. (The Netherlands)	– Need contingencies to initiate action if a contaminated product is discovered – Doing the right thing means ensuring the product you ship is safe, regardless of certificates from upstream suppliers. This means paying attention to your own quality tests – Implement quality and stewardship standards regardless of the end destination for product (i.e., Haiti)
HELM AG (Germany)	– Big company means big responsibility – Need to ensure end user maintains sufficient standards of use for the product, or else refuse to sell – Need to practice more executive oversight and direction as to the principles and values of the company, so that simple profitability is not the sole objective – Need to audit customers and suppliers for suitability of end use and quality of product – Take greater responsibility for ensuring that all customers are satisfied with the product, even those in less-developed nations (i.e., bring Pharval reps to Germany to inspect product, provide testing equipment, education) – Institute an executive position solely responsible for product stewardship and responsible care, whose compensation is based on adherence to these principles
Pharval (Haiti)	– Needs to incorporate quality testing into their manufacturing process, cannot just trust the supposed source – Greater awareness of the actual source of their product – Even if they do not have the resources to ensure the suitability of the product, they can require better stewardship practices of the companies that they buy from – Engage Haitian government to help develop higher standards for imports, etc.

- If all employees had been invested in the company's reputation, then one or both of these issues would not have occurred (the test would have come back before shipping, and the results would have raised a flag).
- These process issues were an underlying problem that could have resulted in tragedy in any number of cases.

– We do not know, however, that what happened in this case really was standard practice (it could have just been a combination of errors).
– There may have been different considerations regarding reputation because they were dealing with a country in Haiti (perhaps they didn't value that business as much as they might have a US company).
– Had they cared more about their reputation, they would have fixed their process issues, and if they were not testing the material before shipping, they would have ensured the product was being tested at the other end (Pharval).
– If not, they would not have sold to Pharval.

Helm AG:
– Probably have the most money at stake in terms of reputation, because they are the largest company.
– That they refused to be accountable for their subsidiary shows that they protect their reputation only superficially (perhaps only "being seen to do the right thing" instead of both doing the right thing and being seen to do the right thing).
– Why did they allow the product to be stamped made in Germany when it was not?
– Under RC, the company would have acknowledged mistakes and would have kept track of how subsidiaries operated.
– Would not have used the legal loophole that since contamination did not occur in Germany they were not responsible.
– Name change of VOS B.V. to Helm Chemicals: Does this signify that, since the child company now shares a name with the parent, they both share a reputation and that reputation is now more important?

Pharval:
– The closest to consumers on the supply chain: probably the most to lose in terms of business based on reputation since they are a smaller company.
– The fact that they did not have testing facilities meant that they were relying too heavily on the reputations of the companies they were dealing with.
– Under RC they would not have stayed in business while operating unsafely.

Concluding thoughts:
– If someone along the supply chain is not acting ethically, RC would have isolated that action.
– How much do investors care about a company's reputation versus profit?
– The longer the supply chain, the more RC seems valuable,
– Ethics and values may differ internationally in the chemical wholesaling business, the reputation is between companies and is not necessarily public (you do not ever see the name Helm AG on a consumer product).

Group 3: *The major players involved did not embed the philosophies of the RC ethic or create a corporate culture of protection for all stakeholders.*

- Transparency: Downstream customers (including other companies, cannot see all the players involved in the supply chain).
- Accountability: All players were unwilling to take responsibility.
- Take preventative actions: Testing was absent at all the stages:
  - Pharval did not have ability/means to test.
  - VOS B.V. tested after shipping and did not forward test results.
  - Sinochem may or may not have tested but did not forward results as product was labeled as pharmaceutical grade.
  - Helm stamped without testing.
- Risk Assessment: Assessing worst case was not done.
- Striving to do no harm: By shipping products that were later found to be toxic was not adhering to this philosophy.
- Innovate for safer products: It seemed as if all players were doing the bare minimum and not looking to do any more than or pushing the bar forward on the product.
- Engage with business partners to ensure the stewardships and security of our products, services, and raw materials throughout their life cycles: No partnerships companies were not aware of other players.
- Understand and meet expectations for social responsibility: Regardless of where the product is going VOS B.V. should have had a level of social responsibility for the end users of their product.
- Work with all stakeholders for public policy and standards that enhance sustainability, act to advance legal requirements, and meet or exceed their letter and spirit: No international legislation, regulation, or standards.
- Promote awareness of RC and inspire others to commit to these principles: No supply chain management, no requirement for suppliers to adhere to any common vision/regulations/plan.
- Continuous improvement: No one was trying to change or improve the system.

Doing the right thing and be seen to do the right thing:

Pharmaceutical companies are in a position to understand the risks and what the "right thing" to do is, far better than the average consumer. The risks they take impact many people and are not often understood by many. Many of the players in this case study did not do the right thing because they were not required to, either by legislation or by consumer/market demands. Likewise there was no communication of the decisions and risks they took.

Doing the right thing and being seen to do the right thing would include:
- Being truthful and stating that the product was only packaged in Germany.
- Being diligent in alerting appropriate authorities about the positive toxic results and recalling the entire batch or affected shipment – thus safeguarding public safety.

- Being accountable – honorable and responsible by assuming responsibility for the deaths or activities leading to deaths.

Major players like BSF did the right thing as prescribed by RC, other smaller pharmaceutical and chemical companies would follow suit. BSF as a huge company would be able to serve as a role model and exert enough clout to encourage others in the industry to embrace RC.

Had the major players involved embedded the philosophies of the RC ethic or created a corporate culture of protection for stakeholders:

- Transparency: Had Pharval known their product was actually made in China, perhaps they would have tested their products before going to market.
- Accountability: Testing would have been more likely if each player had embraced accountability. It would be in their best interest to ensure that products were safe, if they were going to be accountable for it:
    - Chain of custody: Make sure before accepting products the products and documentation are acceptable.
- Risk management: ISO certification/or RC would be one way to minimize risk. Had a risk assessment been done with regards to the Sinochem plant as a supplier they may not have been deemed an acceptable risk, or further testing would have been required.
- Testing would have been more likely had risk assessments been done.
- Work with all stakeholders: Partnerships and cooperation between suppliers and customers could have resulted in better products, testing, accountability, and transparency.
- Promote awareness of RC and inspire others to commit to these – supply chain management (as well as customer management) where preference is given to suppliers that demonstrate adherence to RC principles.
- Continuous improvement: Had any of the companies embraced continuous improvement quality standards would likely have been higher.
- Engage with business partners to ensure the stewardships and security of our products, services and raw materials throughout their life cycles: A potential partnership would have been to collaborate (by Pharval) the means to test products coming into the country.
- Innovate for safer products: Could have involved supplying the customer with the product the means to test and a return system (cradle-to-cradle), looking for substitutes that would be less toxic, indicators to show that the product is still good, etc.

Issues that we see as preventing companies from embedding the RC ethic:
- Costs (initial costs).
- Incentivize RC companies.
- Branding is not very connected to the products.
- RC logo – let people know who the good brands are, not just the bad guys cultural difference (right to know, cradle to cradle, etc. are not common cultural ideals).

Index

www.ingramcontent.com/pod-product-compliance
Lightning Source LLC
Chambersburg PA
CBHW081234190326
41458CB00016B/5777